Developments in Soil Science 1

BAUXITES

Developments in Soil Science 1

BAUXITES

BY

IDA VALETON

Professor of Mineralogy and Petrography
Geological - Palaeontological Institute
University of Hamburg, Germany

ELSEVIER PUBLISHING COMPANY Amsterdam London New York 1972

ELSEVIER PUBLISHING COMPANY
335 JAN VAN GALENSTRAAT, P.O. BOX 211, AMSTERDAM, THE NETHERLANDS

AMERICAN ELSEVIER PUBLISHING COMPANY, INC.
52 VANDERBILT AVENUE, NEW YORK, NEW YORK 10017

LIBRARY OF CONGRESS CARD NUMBER: 70–151740

ISBN 0–444–40888–6

WITH 119 ILLUSTRATIONS AND 42 TABLES

PRINTED IN THE NETHERLANDS

Contents

Chapter 1

INTRODUCTION

This book aims to present an accurate survey of the current state of our knowledge regarding the mineralogy, geochemistry, geology and genesis of bauxite deposits, and to discuss unsolved problems as necessary for teaching and research. For mining engineers and economic geologists the book contributes to a better understanding of the genesis and distribution of different types of deposits.

During the past 10–15 years I have had an opportunity to study various types of deposits both in the field and in the laboratory, but I have also drawn on the comprehensive literature available. Correspondence with experts all over the world not only facilitated an insight into the multiplicity of models of scientific thought, but assisted also in pointing up the complexity of bauxite genesis.

I acknowledge the help of all the scientists and engineers who contributed to our understanding of the principles of bauxite genesis, and wish to thank especially the bauxite companies of France, Greece, Surinam, Guyana and India for all their help during the field work.

I also owe a debt of gratitude to Dr. H. Gebert, who translated and read the entire text. (The original German manuscript was completed in 1968. For this reason latest references are quoted for the year 1968.)

I am grateful to Mrs. A. Berger for her excellent art work and to Mrs. E. Seidlitz for her secretarial help in the preparation of the manuscript.

DEFINITIONS OF BAUXITE, ALUMINA AND ALUMINIUM

Bauxite was the term introduced by BERTHIER (1821) for sediments rich in alumina from the vicinity of Les Baux in the Alpilles (Bouches du Rhône), France. A. Liebrich was the first to extend the term to cover lateritic weathering products rich in gibbsite on basalts of the Vogelsberg in Germany (LIEBRICH, 1892). The term bauxite, therefore, is used for weathering products rich in alumina but low in alkalis, alkaline earths and silica.

The term *bauxite ore* is applied to bauxites which are economically mineable at present or in the foreseeable future, containing not less than 45–50% Al_2O_3 and not more than 20% Fe_2O_3 and 3–5% combined silica.

I shall only discuss ore deposits derived directly from lateritic weathering or from redepositions of such. Alaun schists and ore deposits of non-sedimentary origin, such as cryolites or rocks rich in feldspars and foids, are of minor economic importance and will not be described in this book.

Bauxite deposits originate from weathering or soil formation with enrichments of aluminium. Therefore, principles of red earth formation and lateritization would require discussion in this context. In order to adhere to the scope of this book, the reader is referred to corresponding papers of tropical soil science. Besides aluminium, several elements such as iron, manganese and nickel may enrich to form important ore deposits. A summary of these deposits has been published (VALETON, 1967a).

The following authors surveyed the geology and mineralogy of bauxite deposits and compiled the literature available: FISCHER (1955), SMITH-BRACEWELL (1962), BENESLAVSKY (1963), BARDOSSY (1966), BUCHINSKY (1966) and PATTERSON (1967).

The term *alumina* refers to pure Al_2O_3 which contains 52.9% Al and 47.1% O. The chemical composition of technical alumina is listed in PATTERSON (1967).

The following history of *aluminium* is recorded by GINSBERG (1962):
Herodotus already named alaun or alunite $KAl_3(OH)_6(SO_4)_2$ alumen. In 1754 Marggraf recognized that clay and alaun contain the same cations. A. L. Lavoisier, in 1782, suspected alumina to be the oxide of a metal with such a strong affinity for oxygen that it was not reducible to metal by carbon or any other reductive means. De Morveau (1786) named the hypothetical metal "alumine" (in English alumina). In 1808, Sir Humphrey Davey, who tried to obtain the metal by electrolysis, called it aluminium. In 1824 H. C. Oersted started the preparation of the metal from "water-free" Al_2Cl_3 with potassium amalgam. Continuing OERSTED's (1824) experiments, in 1821 F. Wöhler succeeded in unambiguously isolating the new metal in the form of fine tinsel.

About 1854, H. Sainte-Claire Deville obtained for the first time larger amounts of aluminium from reduction of Al_2O_3 with sodium in a chemical factory near Paris. At the same time, BUNSEN (1852) and SAINTE-CLAIRE DEVILLE (1854) succeeded in producing aluminium by electrolysis. VON SIEMENS (1878) was the first to build a smelter with coal electrodes and coal crucibles. Only after invention of the dynamo in 1866, Heroult in France and, independently of him, Hall in America developed a process for aluminium electrolysis (Hall-Heroult process).

The first electrolysis-based aluminium foundries began production in 1888/1889 in Pittsburgh (Great Britain) and Neuhausen (Switzerland). The technology of aluminium production is intentionally omitted from this book in order to remain within the framework of ore deposit presentation. For further technological orientation the reader is referred to NEWSOME et al. (1960), GINSBERG (1962), SMITH-BRACEWELL (1962), and *Symposium sur les Bauxites, 3, Zagreb, 1963.*

Chapter 2

ECONOMIC IMPORTANCE OF BAUXITE DEPOSITS

UTILIZATION OF BAUXITES

The production of *alumina* consumes over 90% of the bauxite world production (excluding the U.S.S.R. and China), while the remainder is used for *abrasives, chemicals* and *refractories.*

Depending on the mineral composition of the bauxite (gibbsite, boehmite, diaspore) several different processes are used for metal production in the world. The *Bayer process* is the most important process.

The aluminium obtained is used as metallic aluminium or for alloys of: Al–Cu–Mg, Al–Cu–Ni, Al–Cu (5% Cu), Al–Mg–Si, Al–Mg (7% Mg), Al–Mn (2% Mn).

DISTRIBUTION OF BAUXITE DEPOSITS

Throughout the terrestrial epochs of the earth's history bauxites are recorded from all those areas (Fig.5) where tropical weathering resulted in enrichment of aluminium hydroxides and hydrated oxides.

The bauxites may be classified in three different ways (see Fig.6):

on *genetic principles*: (*1*) bauxites on igneous and metamorphic rocks; (*2*) bauxites on sediments: (*a*) carbonate rocks; (*b*) clastic strata;

according to *geological age*: (*1*) Palaeozoic bauxites; (*2*) Mesozoic bauxites; (*3*) Cenozoic bauxites;

based on *mineralogical composition*: (*1*) gibbsite bauxites; (*2*) boehmite bauxites; (*3*) diaspore bauxites.

BAUXITE PRODUCTION

The economic potential of a bauxite deposit is governed by quality, structural geology, reserves, production technology and site in respect to transport facilities.

World production of bauxite is surveyed in Table I. Because of the increasing significance of aluminium for a variety of purposes, production rates rise rapidly.

TABLE I

PRODUCTION OF BAUXITE (in 1000 metric tons)[1]

	1960	1961	1962	1963	1964	1965	1966	1967	1968	1969
Austria	26.0	18.0	17.0	17.8	3.7	—	—	—	—	—
France	2 067.3	2 224.9	2 194.3	2 028.9	2 432.7	2 663.8	2 810.6	2 812.6	2 712.9	2 772.7
Germany (BRD)	3.8	4.2	4.7	4.3	4.2	3.9	3.7	2.3	3.3	3.2
Greece	883.7	1 108.0	1 321.0	1 281.1	1 063.1	1 274.0	1 529.0	1 658.9	1 836.1	1 940.0
Hungary	1 190.0	1 366.0	1 473.0	1 362.0	1 488.0	1 478.0	1 429.3	1 649.4	1 959.0	1 936.0
Italy	315.5	327.2	309.4	269.7	236.1	244.4	253.7	241.4	216.2	228.0
Rumania	88.0	69.0	31.0	10.0	6.6	108.0	206.0	460.0	595.0	600.0
Spain	2.6	5.6	6.0	11.8	6.8	4.2	4.0	2.4	2.9	2.3
U.S.S.R. (including Soviet-Asia)	3 500.0	4 000.0	4 200.0	4 300.0	4 300.0	4 700.0	4 800.0	5 000.0	5 000.0	5 200.0
Yugoslavia	1 025.0	1 232.0	1 332.0	1 285.0	1 293.0	1 574.0	1 887.0	2 131.0	2 072.0	2 128.0
Europe	9 101.9	10 354.9	10 888.4	10 570.6	10 834.2	12 050.3	12 923.3	13 958.0	14 397.4	14 810.2
China	350.0	350.0	400.0	400.0	400.0	400.0	400.0	400.0	400.0	400.0
India	387.4	475.9	573.0	565.1	591.0	705.7	749.8	788.5	936.3	1 012.5
Indonesia	395.7	419.9	461.2	493.1	647.8	688.3	701.5	920.2	879.3	927.0
Malaysia	459.2	416.5	355.0	451.2	471.3	856.7	955.5	899.6	798.7	1 073.0
Pakistan	0.6	1.4	—	—	4.5	10.3	—	—	—	—
Sarawak	289.4	257.5	229.1	157.7	160.6	137.3	32.3	21.0	20.0	20.0
Turkey	—	—	—	—	—	—	—	—	—	—
Asia	1 882.3	1 921.2	2 018.3	2 067.1	2 275.2	2 798.3	2 839.1	3 029.3	3 034.3	3 432.5
Guinea	1 378.0	1 766.7	1 659.8	1 664.0	1 433.0	1 600.0	1 608.7	1 639.2	2 117.6	2 500.0
Ghana	193.8	203.9	243.3	314.4	250.4	319.3	322.9	351.0	284.7	270.0
Mozambique	4.8	4.7	6.2	6.6	6.3	5.7	5.9	5.9	3.3	3.3
Rhodesia	—	—	0.5	1.8	2.5	2.0	2.0	2.0	2.0	2.0
Sierra Leone	—	—	—	30.4	130.5	207.3	275.2	334.5	470.3	442.3
Africa	1 576.6	1 975.3	1 909.8	2 017.2	1 822.7	2 134.3	2 214.7	2 332.6	2 877.9	3 217.6

United States	2 030.1	1 247.7	1 391.0	1 549.5	1 626.7	1 680.5	1 824.8	1 680.5	1 691.7	1 824.8
Brazil	120.8	111.4	190.7	169.6	131.7	188.0	249.9	302.9	313.0	320.0
Dominican Republic	688.6	750.5	675.4	723.5	813.5	893.0	818.4	1 092.0	1 008.0	1 102.8
Guyana	2 510.8	2 411.7	2 762.6	2 163.2	2 517.4	2 918.7	3 357.7	3 381.4	3 721.8	4 306.4
Haiti	346.5	267.2	441.6	384.0	437.2	382.6	361.4	375.8	477.4	776.0
Jamaica	5 837.0	6 770.0	7 615.4	7 013.9	7 936.5	8 651.0	9 061.5	9 395.6	8 525.0	10 498.0
Surinam	3 455.0	3 453.0	3 297.0	3 438.0	3 993.0	4 360.0	5 563.0	5 466.0	5 660.0	6 236.0
America	14 988.8	15 011.5	16 373.7	15 441.7	17 456.0	19 073.8	21 236.7	21 694.2	21 396.9	25 064.0
Australia	70.5	16.2	30.0	359.9	796.5	1 186.4	1 827.1	4 243.6	4 955.0	7 917.0
Total	27 620.1	29 279.1	31 220.2	30 456.5	33 184.6	37 243.1	41 040.9	45 257.7	46 661.5	54 441.3

[1] Metallgesellschaft Frankfurt 1970

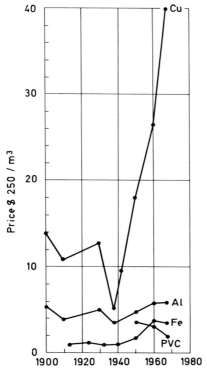

Fig.1. Price development of aluminium, copper, steel and P.V.C. (After BIELFELDT, 1968.)

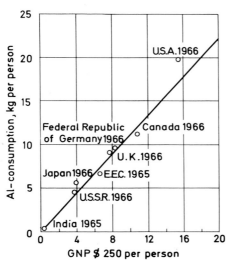

Fig.2. Relation of aluminium consumption and G.N.P. per inhabitant. E.E.C. = European Common Market. (After BIELFELDT, 1968.)

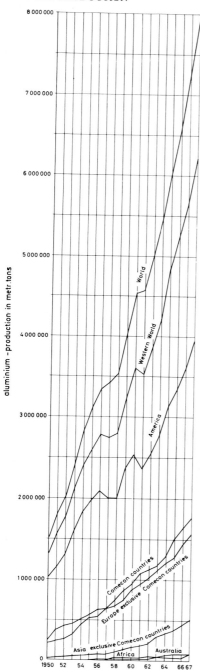

Fig.3. Production rates of aluminium (metric tons; 1 long ton = 1,016 metric tons) in various areas. (After ERNST, 1968.)

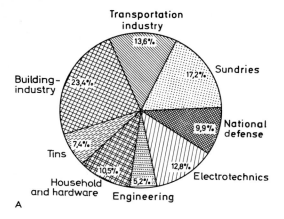

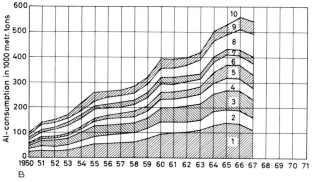

Fig.4. Aluminium consumption. A. In the U.S.A. B. In the Federal Republic of Germany (after ERNST, 1968).

1 = transportation industry; *2* = engineering; *3* = electrotechnics; *4* = packaging; *5* = construction; *6* = furniture and hardware; *7* = steel industry and aluminium powder; *8* = export; *9* = miscellaneous; *10* = summary.

World aluminium production already exceeds the combined output of lead, zinc and tin. Reference to Fig.1 will show that aluminium prices are very stable in contrast to copper, the conductivity of which is two-thirds higher.

The postwar P.V.C. plastics are lower in price but the tensile strength is only 20%–10% of the corresponding value for aluminium. For this reason P.V.C. plastics cannot be substituted for aluminium in many cases. Aluminium satisfies a wide range of industrial needs. Technological advantages of aluminium over other metals are low density, high electric and thermal conductivity, and high imperviosity to corrosion.

It is interesting to compare the production with the rate of importation into the U.S.A. Importation and consumption of aluminium rise sharply with industrialization.

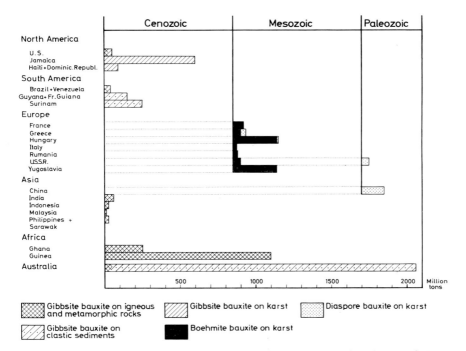

Fig.6. Inferred world bauxite reserves. The most important deposits are of Eocene age. (Bureau of Mines, Washington, 1965.)

World production of aluminium rose from 1.5 million metric tons in 1950 to 8 million metric tons in 1967 (Fig.3). Aluminium consumption reflects the state of industrialization of a country to a large extent and there is a close relationship between G.N.P. (Gross National Product) and aluminium consumption (Fig.2). The United States is by far the greatest aluminium consumer, and consumption

Fig.7. "Mining for bauxite" stamp of Guyana, formerly British Guiana, 1954.

even exceeds expectations. Aluminium consumption in individual branches of industry differs in various countries, and Fig.4 compares commodities of the U.S.A. and the Federal Republic of Germany.

Bauxite production faces the tentatively calculated reserves valued in tons, as given in Fig.6 and in Table I.

In 1954 in Guyana, formerly British Guiana, stamps were issued depicting Her Majesty Queen Elizabeth II and a bauxite mine (Fig.7).

Chapter 3

WEATHERING AND NEOMINERALIZATION

There is a close relationship, as stated by GOLDSCHMIDT (1937), between ionic potential (ionic charge, ionic radius) and the formation of aluminium hydroxides.

By solution in the weathering process some of the trace elements, such as Be^{2+}, Ni^{2+}, Sc^{3+}, Cr^{3+}, Th^{4+}, U^{4+}, Zr^{4+}, Mn^{4+}, Mo^{4+}, Nb^{5+}, V^{5+}, U^{6+}, may be enriched with main elements such as Fe^{3+}, Al^{3+} and Ti^{4+} (Fig.8).

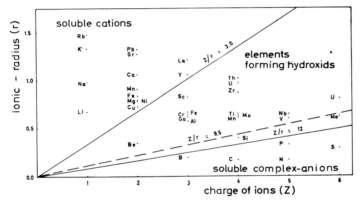

Fig.8. Relation of ionic potential and hydroxide formation. (After GOLDSCHMIDT, 1937.)

MINERAL DECOMPOSITION IN THE WEATHERING PROCESS

Current theories on the disintegration of primary minerals and neomineralization are still very tentative. For this reason and because of the evergrowing number of publications which are partly contradictory, presentation is restricted to selected data and models of thought. There are two principal ways of approaching the problem:

(*1*) Observation of natural processes. The best observations of this kind concern the Hawaiian Islands (see below). However, such observations will remain incomplete because the direct relationship of mineral decomposition and neomineralization on one side and physical and chemical properties of the solution on the other, may never be ascertained. Weathering is such a complex process that it can not be equalled under laboratory conditions.

(*2*) Experimental weathering in the laboratory. Early descriptions of tectosilicates resulted from experimental weathering in suspensions carried out by C. W. Correns and collaborators (Table II).

TABLE II

ARTIFICIAL WEATHERING OF POTASSIUM FELDSPAR IN SUSPENSION (after CORRENS and VON ENGEL-
HARDT, 1938) SHOWING THE MAXIMUM AMOUNT OF K_2O, Al_2O_3 AND SiO_2 DISSOLVED FROM THIS
FELDSPAR IN THE ARTIFICIAL WEATHERING SOLUTION UNDER VARIOUS CONDITIONS OF pH, GRAIN
SIZE, AND RATE OF SOLVENT FLOW

	pH^1				
	3	*3*	*6.6*	*5.6–6.85–8.3*	*11*
Temperature	20	20 °	20 °	20 °	20 °
Grain size, μ	< 1	3–6 (agitated)	< 1	ball mill	< 1
cm³/day	70	~ 74	32	—	32
mg/l K_2O	7.7	11.5	12.3	41.6	14.8
mg/l Al_2O_3	5.6	7	1.1	1	6.5
mg/l SiO_2	10	6.5	7	2	15.8
mg/l solvent	9.1	1.82	1.85	4.5	5.2

[1] pH = 3 by using 0.01 sulphuric acid; pH = 6.6 by using twice distilled water; pH = 11 by using ammonia.

Tectosilicates

CORRENS and VON ENGELHARDT (1938) were the first to demonstrate the dependence of solubility of specific ions on pH. Experiments were carried out with potassium feldspars and decomposition was at its minimum at about pH 7. There is a rapid acceleration in dissolution during initial stages of the experiment, but the speed of dissolution becomes linear from a certain point onwards. Because of the relatively high solubility of potassium, an amorphous residual layer rich in Si and Al 300 Å thick develops. Diffusion generates continuous removal of potassium. Because of this residual layer of constant thickness Si and Al may permeate into ionic solution.

The SiO_2/Al_2O_3 ratio of the residual layer is governed by pH and it is greatest at neutral pH.

Similar experiments were carried out by C. W. Correns and collaborators with plagioclases, foids, amphiboles and olivine. CORRENS (1963) demonstrated the same principles of dissolution for all tectosilicates. In all cases mineral grains develop residual layers. The thickness and chemical composition of these residual layers are governed by the solubility of participating elements under the prevailing conditions.

WOLLAST (1963) reproduced feldspar weathering, both by reactions of aqueous solutions with a feldspar suspension, confirming the results of CORRENS and VON ENGELHARDT (1938), and by diffusion of aqueous solutions in solid phases of pulverized feldspar (100 g/l). In this experiment the reactive surface between solid and solution is much enlarged, and residual layers do not form. Table III compares the solubility of Al_2O_3 and SiO_2 for both methods.

TABLE III

CONCENTRATION OF SILICA AND ALUMINA (mg/l) IN EQUILIBRIUM WITH FELDSPAR (100 g/l) (after WOLLAST, 1963)

pH	Suspension		Percolation		Mean values	
	SiO_2	Al_2O_3	SiO_2	Al_2O_3	SiO_2	Al_2O_3
4	87.8	139	—	—	87.8	139
5	48.9	28.6	39.8	22.2	44.3	25.4
8	17.9	—	17.9	0.65	17.9	0.65
9	15.5	—	16.3	(0.10)	15.9	(0.10)
10	22.5	1.6	—	—	22.5	1.6
H_2O	—	—	33.1	0.45	33.1	0.45
pH final 6.8						

The concentration of Al_2O_3 and SiO_2 in solution after diffusion in equilibrium with feldspar is given in Table IV. Silica is much more soluble than aluminium.

In soils, precipitates from solution form Al/Si ratios governed by both pH and speed of silica removal from solution (WOLLAST, 1963). In similar ways relics, solutions and neomineralization were analyzed in experimental weathering of various rocks by PEDRO (1964).

TABLE IV

COMPARISON OF CONCENTRATIONS OF SiO_2 AND Al_2O_3 IN EQUILIBRIUM WITH FELDSPAR AT DIFFERENT pH (after WOLLAST, 1963)

pH	SiO_2 (millimoles/l)	Al_2O_3 (millimoles/l)
4	1.46	1.35
5	0.74	0.25
H_2O (6,8)	0.55	0.004
8	0.30	0.006
9	0.26	0.0009
10	0.37	0.016

Layer silicates

The decomposition of clay minerals is particularly important in bauxites developed on clayey sediments. Many authors described weathering of layer silicates. WEY and SIEFERT (1961) established increasing stability of mineral phases as follows:

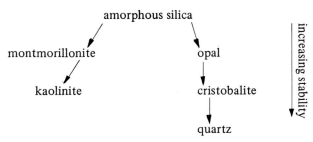

HEYDEMANN (1966) performed a series of experiments on the dissolution of kaolinite, fire clay, illite and montmorillonite at 20°C as a function of pH (pH = 3, 6, 10), which broadened conceptions of possible reactions of clay minerals during weathering.

The investigations consisted of two series of tests. (*1*) A suspension of 1 g of clay in 150 ml of solution was stirred continuously in a filtering apparatus made of plastic material. After 7 days 100 ml of solution were filtered and the remaining 50 ml diluted to 150 ml. This cycle was maintained over a period of 20 weeks. (*2*) Suspensions were shaken in plastic bottles for 1 and 9 months, respectively, at a temperature of 20°C. Disintegration of clay minerals was as listed in Table V.

Kaolinite H-7 dissolves at pH 3 and pH 10 in stoichiometric proportions. However, the total amount in solution is much higher at pH 3 than at pH 10. At about pH 7 decomposition results in residual enrichment of Al. Heydemann suspects stoichiometric dissolution of fire clay (FC) as well, within a pH range of 3–10. He believes that the low concentration of Al depends on the solubility of Al at neutral pH (RAUPACH, 1963) and reflects aluminium absorption on clay-mineral surfaces (RAGLAND and COLEMAN, 1960). There is a smaller dissolution power in fire clay (FC) than in kaolinite in spite of the larger surface of fire clay. This is explained by differences in reaction kinetics.

In the acid range the montmorillonite MU and MP become pH 3 only after several weeks. The final solution after 20 weeks yields a Si/Al ratio of approximately 15 after filtration, but it is approximately 25 after 9 months of shaking. Under acid, neutral and alkaline conditions solubility of silica is relatively high, while the concentration of alumina increases in the residue. High silica enrichments in the filtrate indicate rapid decomposition of the montmorillonite structure. In contrast to kaolinite, disintegration is not stoichiometric but selective with respect to silica.

The transformation of layer silicates into hydroxides has been reproduced

TABLE V

EXPERIMENTAL RESULTS OF ARTIFICIAL WEATHERING (adopted from HEYDEMANN, 1966)

Clay minerals (fraction 2 μ ∅)	pH	After 20 weeks of filtration				After 9 months of shaking			
		concentration in solution		dissolved material		concentration in solution		dissolved material	
		SiO_2	Al_2O_3	SiO_2	Al_2O_3	SiO_2	Al_2O_3	SiO_2	Al_2O_3
		(p.p.m.)		(γ/g clay)		(p.p.m)		(γ/g clay)	
Kaolinite H-7	3	0.62	0.58	1720	1520	6.6	4.7	990	710
1. 45 m²/g clay	6	0.32	<0.02	950	<20	0.8	<0.02	120	<3
2. Si/Al = 1/1	10	0.50	0.46	980	940	0.65	0.70	100	105
Fire clay FC	3	0.60	0.17	1650	420	4.3	1.8	650	270
1. 105 m²/g clay	6	0.08	<0.02	390	<20	1.2	<0.02	180	<5
2. Si/Al = 1/1	10	0.40	0.25	800	570	1.0	0.45	150	70
Montmorillonite MU	3	5.0	0.3	19000	500	28.06	1.1	4300	170
1. 720 m²/g clay	6	1.1	<0.02	5700	<25	7.6	~0.03	1150	~5
2. Si/(Al+Fe+Mg) = 2/1	10	3.0	<0.02	7100	<25	8.0	0.15	1200	20
Montmorillonite MP	3	7.5	0.3	29100	520	41.4	1.7	6210	250
1. 880 m²/g clay	6	1.1	<0.02	7500	<25	~7.0	~0.03	1050	~5
2. Si/(Al+Fe+Mg) = 2/1	10	1.9	<0.02	5100	<25	2.9	0.2	440	30

1 = surface area; 2 = atomic ratio.

by PEDRO and BERRIER (1966) and PEDRO and BITAR (1966) on an experimental basis. They describe both the transformation of kaolinite into boehmite and olivine into brucite via antigorite as the disintegration of the tetrahedron layer and complete structural rearrangement from the amorphous stage after desilification.

Experiments on weathering during the past 30 years evidenced transformation of both tectosilicates and layer silicates into gels via ionic solutions. Within the layer silicate group only, there may be reorganization without structural rearrangement under certain conditions (degradation and aggradation after MILLOT et al., 1966).

Experimental weathering with organic acids

SAPOZHNIKOV (1963) mentions that in recent years the destructive influence of various organic acids upon minerals was studied by I. I. Ginzburg, R. S. Yashina and other I.G.E.M. workers. They used citric, tartaric, pommic, succinic and butyric acids and primary amide of aspartic acid reactions and compared the results with those obtained with sulphuric acid.

The experiments involved the minerals nepheline, chlorite and kaolinite. The concentrations of the acids used ranged from 0.1 N to 0.001 N at room temperature. One gram of the pulverized mineral was placed in a beaker containing 100 ml acid and mixed by stirring for 30 min. Then the mixture was kept for 24 h, and the solution was separated from the precipitate by filtering or centrifuging. The experiment resulted in partial decomposition of all three minerals, even with the solution of lowest concentration. This is easily seen from Table VI.

All mineral extracts contain noticeable amounts of aluminium dissolved by all acids even at their lowest concentration. At 0.05 N all tests reveal considerable quantities of aluminium leached (see Table VII).

TABLE VI

TOTAL AMOUNT DISSOLVED (mg) PER g OF MINERAL (after I. I. Ginzburg and co-workers, in SAPOZHNIKOV, 1963)

Acids	0.05 N			0.001 N		
	nepheline	chlorite	kaolinite[1]	nepheline	chlorite	kaolinite[1]
Citric	304.45	3.96	3.95	2.78	2.32	4.41
Pommic	273.43	5.31	6.10	4.71	1.38	4.80
Tartaric	281.0	3.65	5.57	6.26	1.77	4.34
Aspartic	102.84	4.94	6.95	4.15	1.08	4.26
Butyric	64.95	2.85	4.88	1.22	1.65	4.36
Succinic	73.27	3.12	5.27	3.56	1.41	4.10
Sulphuric	203.17	6.49	5.80	4.24	2.57	4.19

[1] Kaolinite used for the experiment contained some CaO.

TABLE VII

EXTRACTION OF Al_2O_3 PER 1 g OF VARIOUS MINERALS WHEN TREATED WITH ORGANIC ACIDS (0.05 N) (after I. I. Ginzburg and co-workers, in SAPOZHNIKOV, 1963)

Acids	Nepheline		Chlorite		Kaolinite	
	Al_2O_3 (mg/l)	pH of extract	Al_2O_3 (mg/l)	pH of extract	Al_2O_3 (mg/l)	pH of extract
Citric	7.62	6.78	0.50	3.22	0.25	3.50
Pommic	9.43	6.00	0.30	3.12	0.15	3.50
Tartaric	10.70	4.25	0.17	3.12	0.09	3.02
Aspartic	1.50	5.82	0.11	3.52	0.02	3.92
Butyric	0.70	4.22	0.08	3.86	0.01	3.90
Succinic	0.80	4.25	0.07	3.52	0.02	3.88
Sulphuric	3.40	4.20	0.41	2.47	0.30	2.52

To determine the solubility of aluminium with fulvo acids, experiments were carried out by N. V. Saprykina (SAPOZHNIKOV, 1963) in the Caucasian subtropical area, where the laterite type of crust of weathering is known to exist. There was no extraction of aluminium observable upon reaction of fulvo acids with fresh or even gibbsitized rocks. On the contrary, a considerable portion of aluminium present as an admixture in the fulvo acid was absorbed on the rock surfaces (Table VIII).

TABLE VIII

CONTENT OF Al_2O_3 IN SOLUTION AFTER THE FILTRATION OF FULVO ACIDS THROUGH SOIL AND ROCK FOR 24 h (after N. V. Saprykina, in SAPOZHNIKOV, 1963)

| Rock | pH after filtration | Content of Al_2O_3 (mg/l) | | Extraction (−) or absorption (+) of Al_2O_3 |
		in initial fulvo acid	in solution after filtration	
Red soil	5.0	7.54	11.22	3.68 (−)
Hydrargillitized rock	3.68	7.54	0.92	6.62 (+)
Porphyrite	6.93	7.54	5.70	1.84 (+)
Flint clay	3.50	7.54	28.70	21.16 (−)
Bauxite	6.00	7.54	9.94	2.40 (−)

However, filtration of fulvo acids through red soil, flint and bauxite resulted in noticeable extraction of aluminium, the largest amount being extracted from flint clay. There was extraction of aluminium from a rock rich in alumina, e.g., flint clay, if the pulverized rock was left in the fulvo acid solution for a long period

TABLE IX

CONTENT OF Al_2O_3 IN EXTRACTS OBTAINED AS RESULT OF ROCK TREATMENT WITH SOLUTION FOR 30 DAYS (after N. V. Saprykina, in SAPOZHNIKOV, 1963)

| Rock | Weight (g) | Volume of fulvo acid (ml) | pH after rock treatment | Al_2O_3 content (mg/l) | | |
				in initial solution of fulvo acids	in extract	extraction (−) or absorption (+)
Porphyrite	10	150	7.3	7.54	0.00	7.54 (+)
Flint clay	10	150	4.4	7.54	40.99	33.45 (−)
Shingle	10	150	5.0	7.54	0.00	7.54 (+)

(30 days; see Table IX). But within that same period there was no extraction of aluminium from porphyrite and shingle. Moreover, all of it was absorbed from the fulvo acid which initially contained 7.54 mg/l Al_2O_3.

Order of stability of minerals during weathering

Many attempts were made to establish the order of stability of silicates with respect to weathering. As, in the experiment, stability not only depends on the type of mineral (structural bonds) and the environment of weathering, but also on factors such as grain size, inclusions, reacting surface of the mineral and solution (porosity, permeability), no general scheme of mineral stability with respect to weathering may be defined.

PEDRO (1964) diagrammatically demonstrates differences in the speed of dissolution of the tectosilicates from granites and basalts under equal test conditions (Fig.9). Decisive factors governing the speed of degradation seem to be: grain size; greater stability of granite minerals than of basalt minerals with respect to weathering; and higher Si concentration of the granitic weathering solution.

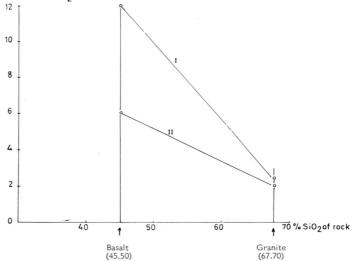

Fig.9. Solubility (wt. %) of silica in basalt and granite as a function of solvent. I = water containing CO_2, pH = 4; II = distilled water, pH = 7. In both cases there is a higher solubility of silica in basalt than in granite. (After PEDRO, 1964.)

The order of stability as derived from weathering profiles on gabbroidic and granitic rocks in the U.S.A. by GOLDICH (1938) may be regarded as one example:

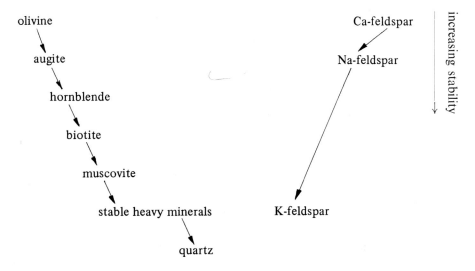

RELIC MINERALS

From both experimental weathering and observations in nature it is now known that minerals disintegrate in a given environment depending on their resistance to weathering. Accordingly they may be stable or instable with respect to the prevailing environment. In principle there are no minerals capable of withstanding tropical weathering.

More stable relic minerals are, for example, quartz, magnetite, ilmenite, chromite, gold, platinum, zircon, rutile, tourmaline. Such relic minerals are of interest: (a) if their concentration is sufficiently high to form residual deposits (chromite, gold, platinum, zircon, rutile); and (b) if they evidence source materials following redeposition (carnotite in karst bauxites of Unterlaussa, Austria).

CHEMICAL REACTIONS IN AQUEOUS SOLUTION AND NEOMINERALIZATION

In bauxites, there is preference to neomineralization of hydroxides, hydrated oxides and oxides of Al, Fe and Ti, but layer silicates and quartz may also form.

Liberation of these elements from a mineral or rock is governed by (1) bonds in the crystal lattice of the minerals to be disintegrated; (2) solubility of the secondary mineral phases; (3) pH and Eh of the solution; (4) charge of the elements, e.g., Fe; (5) temperature and concentration of the weathering solution; and (6) other ions in the weathering solution.

Therefore, the process of dissolution and precipitation of gels or crystals is very complex.

System Al$_2$O$_3$–H$_2$O

HABER (1925) was the first to establish α- and γ-series. Today there are recognized:

    (*1*) α-modifications: Al(OH)$_3$ bayerite
                                AlOOH   diaspore
                                Al$_2$O$_3$ corundum
    (*2*) γ-modifications: Al(OH)$_3$ gibbsite (hydrargillite)
                                AlOOH   boehmite
                                Al$_2$O$_3$

In aqueous solution Al may be dissolved: as ions; in molecular dispersed form; and in polymerized form.

Aluminium is amphoteric in pure water (WEY and SIEFERT, 1961). RAUPACH's (1960, 1963) work on simple aluminium hydroxide–water systems in the pH range 3.5–10, which is of interest in pedogenesis, has shown $AlOH^{2+}$, $Al(OH)_2^+$, and $Al(OH)_4^-$—or possibly polymers: $Al_2(OH)_2^{4+}$, $Al_4(OH)_{10}^{2+}$, $Al_6(OH)_{15}^{3+}$, $Al_4(OH)_8^{4+}$—to be present.

The surface phase of many exchangers, e.g., Al(OH)$_3$, soils and silica, does not liberate Al^{3+} but $AlOH^{2+}$ ions. When Al^{3+} is formed in solution which is slow, they quickly react with the surface conferring a divalent positive charge at the point of reaction and releasing H_3O^+. The results were obtained by dissolving the aluminium hydroxide in 0.01 *M* potassium sulphate (Fig.10). The slope of the line

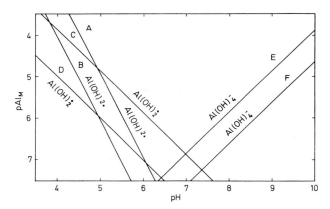

Fig.10. The negative logarithm of the number of g-atoms of soluble aluminium per litre of solution (p Al$_M$) in relation to the pH for various reactions taking place upon dissolution of aluminium hydroxide. The ionic species are indicated; curve *A* and *C* are for amorphous hydroxide, *B* and *D* for corundum, *E* for bayerite, and *F* for gibbsite surface. (After RAUPACH, 1963.)

then gives, when a single aluminium species is present, the number of hydrogen or hydroxyl ions participating in the reaction per aluminium ion, e.g., the line of the equilibrium:

$$Al(OH)_3 = Al(OH)^{2+} + 2 OH^-$$

For precipitated aluminium hydroxide, line A has a slope of 2. The data represent results for pure water systems; lines A and C represent data for freshly precipitated aluminium hydroxide, B and D for corundum, E for bayerite, and F for gibbsite. The free energy of these corresponded to those of stable phases after about 400 h of reaction. While the occurrence of yet more insoluble aluminium hydroxide compounds giving equilibria to the left of lines B and D is quite possible, this would not occur on the alkaline side, where line F for gibbsite is the most soluble phase encountered. RAUPACH (1963) gives a summary of the free-energy values and constants calculated for 25°C (Table X).

TABLE X

FREE-ENERGY VALUES AND EQUILIBRIUM CONSTANTS USED AND EVALUATED[1]

Solid phases	$-\triangle F°$	K_{s0}	K_{s1}	K_{s2}	K_{s4}
Al(OH)$_3$ amorphous	271.9	(−32.34)	−23.31	−14.04	(−10.77)
Al$_2$O$_3$ corundum	273.4*	(−33.45)	−24.41	−15.14	(−11.87)
Al$_2$O$_3$·H$_2$O boehmite	274.2*,**				(−12.45)
Al$_2$O$_3$·3H$_2$O bayerite	276.1*,**				−13.84
Al$_2$O$_3$·3H$_2$O gibbsite	277.1*,**				−14.57
Ions (aqueous)					
Al^{3+}	115.0				
AlOH^{2+}	164.9				
Al(OH)$_2$$^+$	215.1				
Al(OH)$_4$$^-$	313.9**				

[1] K_{s0} = log [(Al^{3+}) (OH$^-$)^3]; K_{s1} = log [(AlOH^{2+}) (OH$^-$)^2]; K_{s2} = log [(Al(OH)$_2$$^+$) (OH$^-$)]; K_{s4} = log [(Al(OH)$_4$$^-$) (H$^+$)].
Values assumed are in italics and those calculated but not found with the present results are in brackets.
* Reduced to formula Al(OH)$_3$.
** In agreement with experimental data of RAUPACH (1963).

Compounds with the free energy of bayerite and gibbsite appear to be the most stable phases on the alkaline side, and hydrated corundum and amorphous aluminium hydroxide in the acid region.

Amorphous stage
Two kinds of gels occur in nature:
1 mole Al$_2$O$_3$ · 3 moles H$_2$O
1 mole Al$_2$O$_3$ · 1.5–2 moles H$_2$O = pseudoboehmite (cliachite)
PA HO HSU and BATES (1964) developed the following experiment based on the precipitation of aluminium hydroxides (Fig.11).

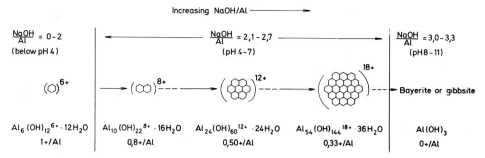

Fig.11. Proposed development of aluminium hydroxides. (After PA HO HSU and BATES, 1964.)

Various crystalline and X-ray–amorphous forms of aluminium hydroxide may be prepared from sulphate and chloride solutions by varying the mole ratio $NaOH/Al^{3+}$. At ratios of 3 or 3.3, crystalline $Al(OH)_3$ as bayerite, nordstrandite, gibbsite, or a mixture is obtained within several hours. At ratios of 2.75 or less no $Al(OH)_3$ is formed, but the products are amorphous to X-raying and remain so even after ageing for 6 months. Chemical analysis indicates basic aluminium sulphates or chlorides for these amorphous precipitates. In the 0–2.1 range of NaOH/Al, the composition of the products is constant, being approximately $Al(OH)_{2.2}X_{0.8}$. In the 2.1–2.75 range, a continuous series of basic salts is obtained, the composition of which ranges from $Al(OH)_{2.2}X_{0.8}$ to $Al(OH)_{2.75}X_{0.25}$.

The following hypothesis is proposed to account for the occurrence of X-ray–amorphous versus crystalline forms. If NaOH is added to an aluminium salt solution, the initial reaction yields $Al(OH)_2^+$, which polymerizes to stable six-membered ring units of $Al_6(OH)_{12}^{6+}$ or multiples thereof (double or triple rings of $Al_{10}(OH)_{22}^{8+}$ and $Al_{13}(OH)_{30}^{9+}$, all with H_2O molecules at the surface. At NaOH/Al = 2, installation of Al^{3+} in the basic ring units is complete; excessive OH^- polymerizes these ring units yielding a continuous series of species. At NaOH/Al = 2.75 and less, the positively charged hydroxide-aluminium polymers repel each other unless joined together by anions to form basic aluminium salts, which are usually amorphous to X-rays, because of their highly hydrated state. At NaOH/Al = 3, the net positive charge per aluminium is neutralized and consequently repulsion among both polymers and hydrated compounds becomes negligible. For this reason all polymers cluster, and $Al(OH)_3$ crystallizes within hours or days, depending on the conditions of crystal growth (Fig.11).

Trihydrate $Al(OH)_3$ (α = bayerite, γ = gibbsite (hydrargillite), nordstrandite)

All $Al(OH)_3$ modifications (Table XI) have similar crystal structures, showing

OH octahedrons in which two thirds of the centres are occupied by Al (dioctaedric) linked to layers with pseudohexagonal grouping of Al. The various polymorphous types are defined by the different structural arrangement of successive layers. The structure of bayerite is very near to hexagonal "closed sphere packing" resulting in higher density than that of gibbsite, layers of which are displaced along the *a*-axis to such a degree that they face OH-ions of neighbouring layers.

SAALFELD (1960) describes monoclinic and triclinic gibbsite which are distinguished from structural arrangement of successive layers along the *c*-axis.

GINSBERG et al. (1962) proved bayerite to be the most stable mineral phase in an environment free of alkalis.

The normal case is that in the presence of alkalis the gibbsite structure forms. The alkalis are built into the lattice, the amount ranging from 0.1 % to 1 %. They are likely to fill vacancies in the octahedrons. Gibbsite, therefore, is not pure $Al(OH)_3$ but a modification stabilized by alkalis. HAUSCHILD (1964) was able to stabilize gibbsite with ethanolamine in the absence of alkalis.

In the presence of potassium, WEFERS (1962) grew elongated pseudo-hexagonal gibbsite prisms, while tabular crystals, frequently twisted, developed from the sodium solution. Also large natural crystals in bauxites of India (VALE-

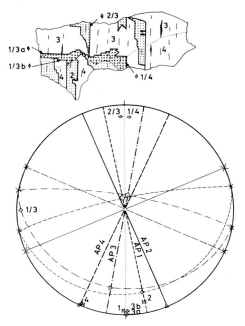

Fig.12. Stereo projection of four individual gibbsite crystals with different triclinity. Crystals *2–3* and *1–4* are twinned after (001) and crystal *1–3* after (010). Note the rotation of crystals *1* and *3* in relation to crystals *2* and *4*, respectively, around a spiral axis [001]. (After VALETON, 1967b).

TON, 1967b) show spiral growth besides twinning in (001) and (010) planes (Fig.12).

The most common naturally occurring trihydrate is gibbsite (= hydrargillite). It crystallizes from both gels (via solution), sometimes in a pseudomorphous type, and directly from solution to form big crystals in joints and rock cavities.

It is the main mineral of nearly all laterite bauxites and many Tertiary karst bauxites (Jamaica, Haiti, southern Europe). Nordstrandite (WALL et al., 1962; HATHAWAY and SCHLANGER, 1965; SAALFELD and MEHROTRA, 1966) and bayerite (GROSS and HELLER, 1963) also became well-known examples of neomineralization. From the white bauxite of Carev Most near Niksic (Montenegro), KARSULIN (1963) describes monoclinic aluminium hydroxides with the formula $2Al_2(OH)_6 . H_2O$ (for data, see Table XI) which he named "tucanite".

Monohydrates Al OOH (α = diaspore, γ = boehmite)

The structural relations of gibbsite and boehmite and diaspore may be read from Fig.13. WYART et al. (1966) investigated the transformation albite → boehmite in experiments. They observed development of corrosion on (001)

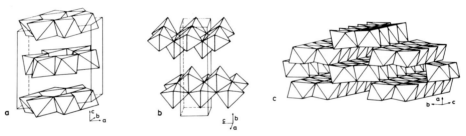

Fig.13. Crystal structures of gibbsite (MEGAW, 1934), boehmite (EWING, 1935) and diaspore (EWING, 1935).

faces during decomposition of albite by water and by compressed CO_2 in the autoclave. The aluminium dissolved immediately precipitated as orientated boehmite fibres which were replaced by tabular boehmite crystals after a few hours.

In natural occurrences boehmite and diaspore are widespread.

In nature, boehmite, which is thermodynamically the unstable phase, predominates over diaspore. It occurs in various types of ore deposits:

(*a*) The pisolitic high-quality bauxite facies of many laterites is rich in boehmite. In places boehmite is even the dominant aluminium mineral.

(*b*) In karst bauxites boehmite may be the predominate aluminium mineral covering Palaeozoic to Tertiary deposits. The crystals both in the matrix and in oolites and pisolites are submicroscopic in most cases.

DE LAPPARENT (1930) describes crystals bounded by (001) and (110) faces from bauxites of the Ariège. Angles of 60° of the (110) facies arise from preferred growth of the basis (001). Twinning after (001) is common.

In nature, diaspore forms both from recrystallization at the expense of gibbsite and boehmite and from solution as big crystals in joints and cavities. Diaspore crystals are much bigger than boehmite crystals (up to 100 μ) in most cases. The habit is pseudomorphous or euhedral with faces (001), (100) and (101).

De Lapparent (1930) illustrates twinning after (031). The former opinion that diaspore formation was possible only at elevated temperatures and pressures, proved to be wrong. Allen (1935) was the first to describe diaspore formation in clays of Missouri (Pennsylvanian) with no signs of metamorphism. In 1952 he described the mineralogical relations of gibbsite, boehmite and diaspore in the diasporiferous clays of Missouri. The conception of diagenetic gibbsite, boehmite and diaspore facies laterally and vertically interfingering was described from several thoroughly investigated areas in Hungary (Bardossy, 1958) and in France (Valeton, 1964).

The diagenetic environment governs "stabilization" of each mineral phase. Both Bardossy and Valeton suspect the oxidation potential to influence mineral formation. Nia (1968) demonstrates replacement of boehmite by diaspore with increasing drainage in bauxites of Greece.

Oxide Al₂O₃

The oxide Al_2O_3 may be represented by several polymorphous types, the crystallographic properties of which are still unknown in part. The stable end product is always α-Al_2O_3–corundum.

Authogenetic corundum was described first by Terentieva (1958) from North Kasaktanian and later by Beneslavsky (1959, 1963) from Russian non-metamorphic Mesozoic and Cenozoic bauxites. Zambo and Toth (1961) found 2–6% α-Al_2O_3 in bauxites of Hungary. It mainly formed in black hard concretions and remained of submicroscopic size.

Rehydration of Al₂O₃

Many experiments (references in Newsome et al., 1960) showed that re-hydration of α-Al_2O_3 to γ-monohydrate (boehmite) is possible at 100°C without difficulty. Further rehydration leads to α- or γ-trihydrate. In some cases lattice transformation lagged behind water resorption. Experiments with higher temperatures and atmospheric pressure or in a vacuum showed rehydration to be more difficult. Beneslavsky (1963) describes natural rehydration from Russian bauxite deposits.

Relations of the mineral phases in the system Al₂O₃–H₂O

So far there is no satisfactory experimentally based explanation of the conditions of mineral formation in the system Al_2O_3–H_2O. For this reason several phase diagrams exist (Fig.14) such as those of Erwin and Osborn (1959), Ken-

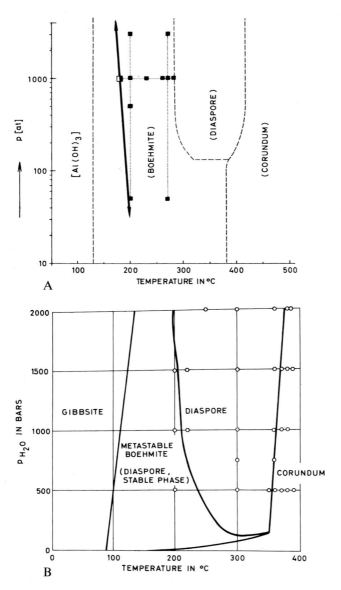

Fig.14. Phase diagrams $Al_2O_3-H_2O$.

A. After Erwin and Osborn (1951) with projected points of diaspore stability after injection of diaspore nuclei (Neuhaus and Heide, 1965).

B. After Kennedy (1959); according to him there is no boehmite stability field.

C. After Torkar and Krischner (1963) with new metastable "Übergangstonerden": α-Al_2O_3; γ-Al_2O_3 (obtained in the autoclave), Al_2O_3 K I.

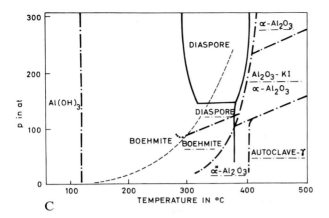

C

NEDY (1959), TORKAR (1960) and TORKAR and KRISCHNER (1963) and NEUHAUS and HEIDE (1965).

According to TORKAR (1960) and TORKAR and KRISCHNER (1963), in the genuine Al_2O_3–H_2O system only bayerite is formed while gibbsite results from the presence of Na^+ and several other impurities. For this reason bayerite must be considered to be the stable $Al(OH)_3$ form, while gibbsite and nordstrandite are stabilized by impurities.

If water vapour pressure is raised to more than 100 atm., both boehmite and diaspore crystallize. It is not yet certain whether both minerals are thermodynamically stable or whether diaspore is the only stable phase, as KENNEDY (1959) assumes. Transitions to α-Al_2O_3 are considered to be non-stoichiometric hydroxides with respect to the amount of OH groups.

The γ-Al_2O_3 group is made of various stages of crystallinity. α-Al_2O_3 is always the stable end product.

Rapid dehydration at the surface of larger grains may develop excessive H_2O pressure in the centre of the grains because of isolation ("Krustentheorie" of B. Frisch), which allows several phases to crystallize within one grain. In order to study changes in crystal structures during phase transformations, modern research follows changes of electron diffraction patterns during dehydration (BRINDLEY, 1961; LIPPENS and DE BOER, 1964). By this means SAALFELD and MEHROTRA (1965) observed orientated transformations (topotaxy) boehmite $\rightarrow$ γ-Al_2O_3.

Additional reflexes are recorded during the transformation boehmite $\rightarrow$ γ-Al_2O_3 which fit the reciprocal lattice of boehmite. Diffuse streaks along c^+ initiate the topotactic transformation into γ-Al_2O_3, frequently being pseudomorphous. The transformation occurs in two ways depending on the crystal size, rate of heating, and furnace temperature:

(1) gibbsite $\xrightarrow{300\,°C}$ boehmite $\xrightarrow{450}$ γ $\xrightarrow{700-900}$ δ $\xrightarrow{900-1000}$ ϑ $\xrightarrow{1200}$ α-Al_2O_3

(2) gibbsite $\xrightarrow{200-300}$ χ $\xrightarrow{800}$ $\varkappa$ $\xrightarrow{1200}$ α-Al_2O_3

NEUHAUS and HEIDE (1965) who refer to highly inhibited reaction systems, tried to eliminate both disintegration and processes inhibiting nuclei formation by innoculation of diaspore nuclei. The lattice of boehmite and diaspore differ significantly, blocking the transformation of boehmite to diaspore. For this reason the transformation of boehmite to diaspore progresses via solution, whereby the dissolution of boehmite is retarded, particularly at low temperatures and low pressures. But there is a genuine equilibrium curve of two stable phases in the diaspore–corundum system, because of good pseudo-three-dimensional topotaxy. This transformation may take place at very low temperatures. The transformation of boehmite into corundum is via solution and the intermediate diaspore phase.

System Al_2O_3–SiO_2–H_2O

Layer silicates of bauxite deposits are minerals of the kaolinite-, montmorillonite- and chlorite groups.

Silica

Silica dissolves fastest and most easily in tropical climates and hence requires a close study.

In water, silica may: (a) polymerize to colloidal silica; (b) form dispersed molecules of $Si(OH)_4$; (c) occur as ions.

The solubility of monomeric silica in undersaturated solutions is a function of: (a) bonds in the crystal lattice; (b) pH; (c) temperature; (d) other ions; (e) concentration of the solution.

At defined temperatures, e.g., 22 °C, there is a strong pH dependency of $Si(OH)_4$ solubility (KRAUSKOPF, 1956, 1959). Below pH = 9 some 120 p.p.m. Si/l are dissolved, and only above pH = 9 does dissociation of $Si(OH)_4$ increase rapidly (Fig.15).

The solubility of silica is strongly dependent on temperatures, too, as demonstrated by OKAMOTO et al. (1957). A rise in temperature from 0° to 73 °C causes the solubility of $Si(OH)_4$ to increase by a factor of 4. Okamoto et al. proved that other ions influence the solubility of silica. The solubility of colloidal and molecular dispersed silica decreases with increasing Al^{3+} concentration (Fig.16, 17).

Salts, e.g., NaCl, increase the adsorption of silica on aluminium hydroxides.

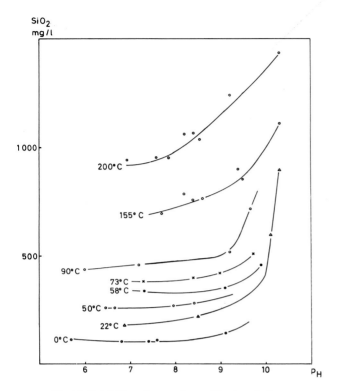

Fig.15. Solubility of amorphous silica in relation to temperature and pH. Rising temperature is paralleled by increasing solubility of silica (OKAMOTO et al., 1957).

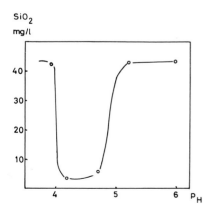

Fig.16. Ions other than silica (e.g., Al) cause the solubility of silica to decrease substantially. $SiO_2 = 45$ mg/l, Al = 1 mg/l (OKAMOTO et al., 1957).

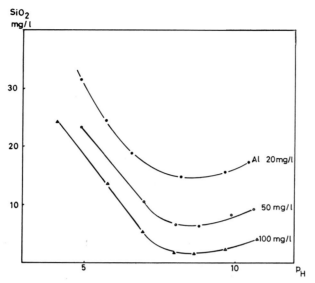

Fig.17. Decreasing solubility of molecular dispersed silica in relation to Al concentration, $SiO_2 = 35$ mg/l (OKAMOTO et al., 1957).

According to HARDER (1965), silica adsorption on aluminium hydroxides is also a function of the concentration of the solution, and the maturity and crystallinity of the aluminium hydroxide gels.

Increasing dilution with a constant Al/Si ratio results in a poor adsorption of silica while excess aluminium precipitates silica quantitatively. There is a large amount of adsorption of silica on highly dispersed aluminium hydroxides, but very little adsorption occurs on aged and well-crystallized phases (refer to WOLLAST, 1963). Increasing temperatures cause silica adsorption on aluminium hydroxides to decrease.

Iron hydroxides behave in much the same way as aluminium hydroxides. Therefore, alkalinity, higher temperatures and dilute solutions favour transport and migration of silica.

Layer silicates

Layer silicates of bauxite deposits may be: (*a*) relic minerals; (*b*) desilicification of three-layer minerals; (*c*) neomineralization by resilicification: reaction of silica solution with amorphous Al-hydroxides.

Relic clay minerals

Minerals of the kaolinite group proved to be more stable than three-layer minerals, as previously shown. Two-layer minerals, therefore, enrich relatively during lateritic weathering.

Desilicification of clay minerals

BATES (1962) observed halloysite → alumina gel → gibbsite transitions in weathering andesites on Hawaii. Further examples of corresponding desilicification were described by KELLER (1964). In soils, the transition of montmorillonite to 14-Å minerals was first observed by BROWN (1953) and later by TAMURA (1958), and SAWHNEY (1958), followed by many other authors. They believe that more or less complete Al-hydroxide layers are introduced in between structural units (= "chloritization of montmorillonite").

ALTSCHULER et al. (1963) described weathering of montmorillonite in a kaolinite deposit in Florida. Based on chemical, X-ray and electron-microscopic investigations, the authors assumed an initial dissolution of exchangeable cations from the octahedron layer of the montmorillonite in ground water of low acidity (pH 6.5–5.0). This is followed by the slow disintegration of tetrahedrons, thus transforming montmorillonite into mixed layers of montmorillonite-kaolinite in a 1/1 proportion.

Altschuler's conception of the formation of defined mixed layers yielding 14-Å reflexes from montmorillonite degradation by selective disintegration of tetrahedron layers seems to be more reasonable than the model of "chloritization" by the introduction of Al-hydroxide layers. The detailed process of degradation and transformation of the clay minerals remains to be clarified by future investigations.

Neomineralization of clay minerals

Neomineralization of clay minerals initiated by the reaction of silica and amorphous aluminium hydroxides is governed by the Al/Si ratios in such gels. The faster solutions are removed by drainage, the less silica is available for reaction with aluminium hydroxides. Accordingly there is a close relationship between speed of drainage and type of clay-mineral neomineralization. According to WOLLAST (1963), the solubility of amorphous aluminium hydroxides exceeds the corresponding gibbsite values by a factor of 10^4. Therefore, clay minerals will normally form from the reaction of silica with amorphous aluminium hydroxides. The order of clay-mineral neomineralization with decreasing silica removal is:

feldspar → gibbsite:
$$2 H^+ + 6 SiO_2 \cdot Al_2O_3 \cdot K_2O + 14 H_2O =$$
$$6 SiO_4H_4 + 2 K^+ + Al_2O_3 \cdot 3 H_2O \tag{1}$$

feldspar → kaolinite:
$$2 H^+ + 6 SiO_2 \cdot Al_2O_3 \cdot K_2O + 9 H_2O =$$
$$4 SiO_4H_4 + 2 K^+ + 2 SiO_2 \cdot Al_2O_3 \cdot 2 H_2O \tag{2}$$

kaolinite → gibbsite:

$$2\ SiO_2 \cdot Al_2O_3 \cdot 2\ H_2O + 5\ H_2O =$$
$$2\ SiO_4H_4 + Al_2O_3 \cdot 3\ H_2O \qquad\qquad\qquad (3)$$

This reaction is reversible, a function of H_4SiO_4 concentration. GARRELS and CHRIST (1965) calculated the equilibrium of K-feldspar and kaolinite on the one hand and kaolinite and gibbsite on the other (Fig.18–20). In nature, there may

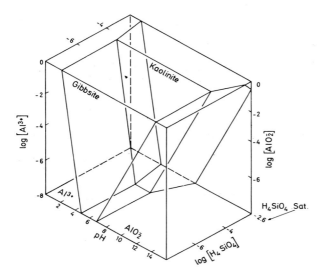

Fig.18. Gibbsite–kaolinite stability fields expressed in terms of pH and activities of the dissociation products Al^{3+}, AlO_2 and H_4SiO_4 at 25 °C and 1 atmosphere. (After GARRELS and CHRIST, 1965.)

be a high proportion of amorphous material. According to HENDRICKS et al. (1967) in the Cascade Range in northeastern California over 30% of the halloysite allophane and sesquioxide allophane formed from andesite weathering to saprolite. For instance, if there is a Al/Si ratio in gels corresponding to kaolinite composition, the term prokaoline is used. Depending on the local supply of ions various clay minerals such as montmorillonite, chlorite, vermiculite and mixed layers are formed. In the upper parts of the profiles of both laterite bauxites and karst-bauxites there may be syngenetic enrichment with SiO_2 and clay-mineral neo-mineralization mostly of the kaolinite groups.

In bauxites, kaolinite or halloysite are by far the most common clay minerals. Kaolinite crystals form both in the matrix and in concretions and pisolites. Kaolinite formation caused by later resilicification of aluminium minerals is subordinate in bauxite deposits. In joints and cavities and on boundaries with

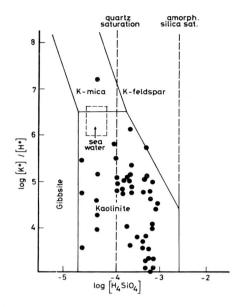

Fig.19. Stability relationships of some phases in the system $K_2O–Al_2O_3–SiO_2–H_2O$ at 25 °C and 1 atmosphere, as functions of $[K^+]/[H^+]$ and $[H_4SiO_4]$. Solid circles represent analyses of waters in arkosic sediments. (After GARRELS and CHRIST, 1965.)

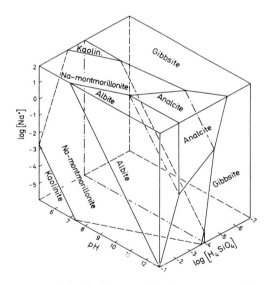

Fig.20. Stability relationships of some phases in the system $Na_2O–Al_2O_3–SiO_2–H_2O$ at 25 °C and 1 atmosphere, as functions of $[Na^+]$, pH and $[H_4SiO_4]$. The diagram shown is an estimate based on extrapolation of experimental relationships at higher temperatures and pressures, and the numerical values are only approximated. (After K. Linn, in GARRELS and CHRIST, 1965.)

neighbouring rocks secondary kaolinite films formed or impregnation developed. Commonly these kaolinites are characterized by a high degree of cristallinity.

Based on many observations the following transformations may be assumed to occur in nature:

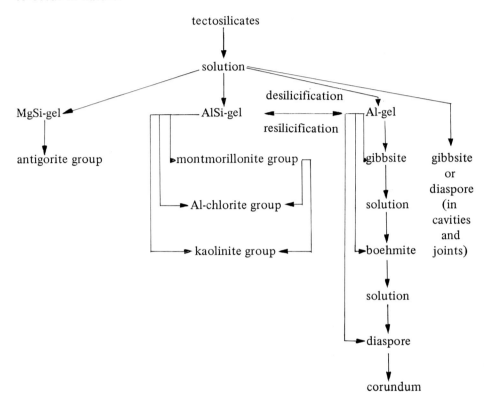

System Fe₂O₃–H₂O

The natural iron compounds are: (*a*) amorphous hydroxides and humic iron complexes of various chemical composition; (*b*) hydrated oxides: α-Fe OOH (goethite), γ-Fe OOH (lepidocrite); (*c*) oxides: α-Fe₂O₃ (hematite), γ-Fe₂O₃ (maghemite), Fe₃O₄ (magnetite).

Discrepancies found between the results of POSJNAK and MERWIN (1922), SCHWIERSCH (1933), SMITH and KIDD (1949), SCHWERTMANN (1959), SCHMALZ (1959), LIMA DE FARIA (1963) and HILLER (1964) show that phase boundaries in the system H₂O–Fe₂O₃ are not yet satisfactorily defined.

This book is concerned only with partial processes of iron precipitation and crystallization. Both Fe²⁺ and Fe³⁺ compounds exist in nature. The ions in solution are both monomeric and polymeric: Fe^{2+}, $[FeOH]^+$, $[FeO_2H]^-$, Fe^{3+}, $[FeOH]^{2+}$, $[Fe(OH)_2]^+$, $[Fe_2(OH)_2]^{4+}$ and $[Fe_4(OH)_4]^-$. There may be precipitation of complex ferriferro hydroxides such as $Fe_3(OH)_8$ and $Fe_4(OH)_{10}$ if Fe^{2+} and Fe^{3+} are in solution. In principle Fe^{3+} compounds may crystallize: (a) directly from solution, by oxidation of Fe^{2+} solution, by hydrolyzation of Fe^{3+} compounds, or by oxidation of Fe^{3+} organic complexes; (b) from amorphous phases. The factors governing crystallization of specific phases are: Eh, pH, concentration of the solution, other ions in solution, ageing conditions of gels, and temperature, among others. SCHELLMANN (1969) calculated the solubility of Fe^{2+} in terms of pH and Eh conditions (Fig.21).

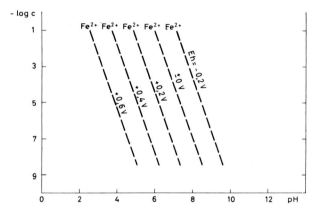

Fig.21. Solubility of Fe^{2+} calculated in relation to pH and Eh. (After SCHELLMANN, 1969.)

The crystallization of α-FeOOH, γ-FeOOH and α-Fe$_2$O$_3$ strongly depends on pH. FEITKNECHT and MICHAELIS (1962) investigated the reaction products formed when $[Fe(H_2O)_6]^{3+}$ is deprotonated in perchlorate solutions. If a small amount of base is added, metastable solutions containing the ions $[Fe(H_2O)]^{3+}$, $[Fe(H_2O)_5OH]^{2+}$, $[Fe_2(H_2O)_8(OH)]^{4+}$ and $[Fe(H_2O)_4(OH)]^+$ are formed, and α-FeOOH slowly crystallizes from solution. In a narrow intermediate range immediately after adding the base, colloidal amorphous Fe(OH)$_3$ and crystalline α-FeOOH are formed. With additional base introduced, a dark brown solution of colloidal amorphous Fe(OH)$_3$ forms. There is crystallization of α-FeOOH and under certain conditions γ-FeOOH may form, too, after an induction period of several days. Only after a nearly equivalent amount of base was added amorphous Fe(OH)$_3$ precipitated. On ageing, this compound changed partly into α-Fe$_2$O$_3$, partly into γ-FeOOH, and a rather large amount remained amorphous.

In all these systems, no stable stage is reached at room temperature, even after years. If small amounts of Fe^{2+} are co-precipitated, a complete crystallization of the amorphous hydroxide to α-FeOOH takes place.

According to SCHINDLER et al. (1963) there is a sediment of active amorphous Fe^{3+} hydroxides and small amounts of fine crystalline α-FeOOH after 200 h of ageing. Further ageing of the Fe^{3+} hydroxides results in amorphous "inactive" hydroxide of $[FeO_{n/2}(OH)_{3n-m}]$ composition and crystalline phases such as α-FeOOH and α-Fe_2O_3. From the results of precipitation and dissolution reactions phase equilibrium data were calculated, and a solubility diagram for amorphous active $Fe(OH)_3$, amorphous inactive $Fe(OH)_3$ and α-FeOOH was constructed (Fig.22). The solubility of aged iron (III) hydroxide precipitates was

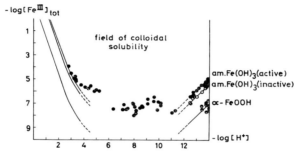

Fig.22. Solubility diagrams of amorphous active $Fe(OH)_3$, amorphous inactive $Fe(OH)_3$ and α-FeOOH.
The points plotted (LENGWEILER et al., 1961) correspond to direct determinations of am. act. $Fe(OH)_3$ (●), am. inact. $Fe(OH)_3$ (○), α-FeOOH (◑); (SCHINDLER et al., 1963).

investigated by determining $[Fe^{3+}]$ and $[H^+]$ of solutions in contact with the solid phases. $[Fe^{3+}]$ and $[H^+]$ were measured by the E.M.F. method at a constant ionic strength of 3M $(Na)ClO_4$. The following equilibrium constants were derived:

$$\log [Fe^{3+}] [H^+]^{-3} = \log Ks = 3.55 \pm 0.1 \text{ (am. inact. hydroxide; 25°C)} = 1.4 \pm 0.8° (\alpha\text{-FeOOH; 25°C)}.$$

By comparing these data with the measurements from BIEDERMANN and SCHINDLER (1957), the following free enthalpies have been calculated:

am. act. $Fe(OH)_3$ = am. inact. $Fe(OH)_3$
       $K_1 = -0.56 \pm 0.2$ kcal (25°C)
am. act. $Fe(OH)_3$ = α-FeOOH + H_2O
       $K_2 = -3.5 \pm 1.1$ kcal (25°C)
am. inact. $Fe(OH)_3$ = α-FeOOH + H_2O
       $K_3 = -2.9 \pm 1.1$ kcal (25°C)

From altered gels at pH 10 there was crystallization of goethite only, whereas below pH 10, hematite also occurred (SCHWERTMANN, 1966). Amorphous hydroxide occurred as spherical particles which are considered to be pre-stages of hematite. Goethite develops in two steps by hydrolysis of Fe^{3+} solutions, whereas hematite forms from dehydration of amorphous Fe^{3+} hydroxides. The overall scheme is:

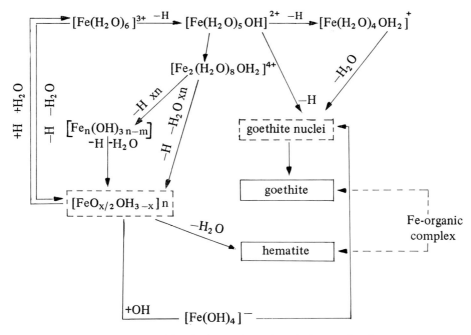

There may be some dissolution of aged and fine-grained amorphous hydroxide that gives rise to goethite crystallization from solution. Therefore, besides other factors, the solubility of the amorphous hydroxide governs goethite and/or hematite formation. In the presence of Fe^{2+}, hematite is easily destroyed and goethite crystallizes out of the solution.

The various ions contaminating the gels may influence both speed and direction of the ageing process. The crystallization is prolonged not only by inorganic anions such as phosphate and silicate but also by organic components. According to the mechanism of formation mentioned above, organic components may also influence the direction of ageing by forming organic metal complexes, thus preventing precipitation of the hydroxide. The transformation goethite $\rightarrow$ hematite is accelerated by higher temperatures.

WEFERS (1966) investigated the goethite-hematite phase boundary (Fig.23) in *hydrothermal solution* experiments. The strong environmental influence upon goethite or hematite formation is expressed by the equilibrium equation:

$$2\ FeOOH \rightleftharpoons Fe_2O_3 + H_2O$$

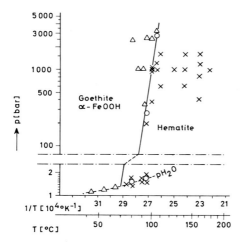

Fig.23. Phase diagram of Fe₂O₃–H₂O.
$\times$ = hematite crystallizes from solution; $\triangle$ = goethite crystallizes from solution; $\circ$ = direction of crystallization not determinable (pH₂O has been added to 1 bar). (After WEFERS, 1966.)

existing at 70°C and water vapour pressure saturation. Only a minor inhibition of the reaction may prevent equilibrium. The common co-existence of goethite and hematite in nature may be explained by this fact.

The conditions of stability of γ-FeOOH (lepidocrocite) and α-FeOOH (goethite) were also revised by WEFERS (1966). Between 25° and 100°C γ-FeOOH is transformed into the more stable α-FeOOH form via solution.

In nature α-minerals are more common than γ-phases. In most cases they are poorly cristalline, and commonly these minerals occur together with amorphous iron hydroxides. The main minerals are goethite and hematite. It is not clear whether hematite or goethite is the primary mineral. But there is no question of the fact that hematite is the primary mineral in saprolites and lateritic iron crusts of laterite bauxite in India. During polygenetic recrystallization hematite is formed in all areas with good drainage, mostly through the separation of Al and Fe as bigger crystals (5–100 μ). In the lateritic iron crust with clayey decomposition hematite changes into goethite via amorphous stages, beginning at the surface of the lateritic profiles.

Polygenetic transformations may result in a variety of iron minerals: BALKAY and BARDOSSY (1967) describe a profile in Guinea with hematite in the upper and maghemite in the lower part, respectively, in addition to goethite. BONIFAS and LEGOUX (1957) record maghemite (γ-Fe₂O₃) from Conary. SCHELL-MANN (1964) also found maghemite (up to 15%) in laterites on serpentinite, but in no case it was associated with gibbsite. Below the lignite cover pyrite, marcasite and siderite replace iron oxides and hydroxides.

System Al_2O_3–Fe_2O_3–H_2O

Isomorphous mixability of goethite and diaspore was first pointed out by CORRENS and VON ENGELHARDT (1941). The lattices of α-FeOOH and α-AlOOH are isomorphous. NORRISH and TAYLOR (1961) investigated natural goethite in Australian soils by X-ray techniques and found that it contained 25 mole % AlOOH (reduced d-values of (111)). There are descriptions of isomorphous replacement of 2–5% Al in goethite in bauxites of India (VALETON, 1966) and Guinea (BALKAY and BARDOSSY, 1967). THIEL (1963) succeeded in experimentally preparing goethite–diaspore mixed crystals with the formula α-$Fe_{0.67}Al_{0.33}OOH$, demonstrating a clear relationship between isomorphous aluminium and lattice constant.

CAILLÈRE (1962) records isomorphous replacement of Al by Fe in boehmite and diaspore in bauxites of France. There are also descriptions of diaspore which may contain isomorphous admixtures of Fe (up to 7 mole %), Mn, Cr, Si and Ga by BETECHTIN (1964).

Because of the fine grain sizes of bauxite minerals it is commonly difficult to distinguish between genuine mixed crystals by isomorphous replacement and mixed crystals resulting from intergrowth. Therefore WEFERS (1967) experimentally prepared the system Al_2O_3–Fe_2O_3–H_2O (Fig.24), and determined by crystallographic means the phases precipitated.

The hydrated iron oxides may store up to several mole % of Al in their lattice (= *MK* in Fig.24). No storage of Fe was recorded in the aluminium compounds gibbsite and diaspore. However, there is some growth of mixed crystals in the α-Fe_2O_3 and α-Al_2O_3 systems, the miscibility gap being from 6 to 96 mole % Al_2O_3. The storage of Fe in the corundum lattice lowers the critical temperature of the diaspore–corundum transformation from 360°C to 330°C at 97 mole % Al_2O_3. Likewise, there is a lower critical temperature of hematite–goethite transformation if Al is stored in the hematite lattice.

Wefers discovered that diaspore crystallized from mixtures of 40–60 mole % Fe_2O_3 at just under 100°C. He believes that the goethite formed during the early stages served as the "crystal nucleus for epitactic growth of diaspore, nearly without stress", as differences in goethite and diaspore lattices are minor. Because of the orientated growth of goethite optimum energy is released by diaspore nuclei formation. The broadly shaped X-ray peaks at low temperatures of closely intergrown diaspore and goethite gradually split with rising temperatures. Dehydration of goethite to hematite and recrystallization of diaspore are contemporaneous.

In the field of low iron content, gibbsite is the primary mineral; it later dehydrates to boehmite in the presence of small amounts of alkalis. The transformation of gibbsite into boehmite may be interpreted as the merging of two $Al(OH)_3$ layers into one Al OOH octahedron double-layer. As vital structural units of the

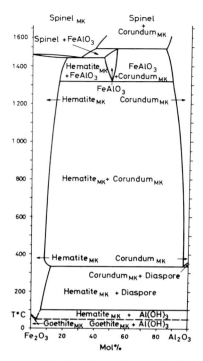

Fig.24. Phase diagram Fe_2O_3–Al_2O_3–H_2O; pseudobinary presentation (related to mole percent of oxides, pH_2O of saturated solutions). (After WEFERS, 1967.)

gibbsite structure are maintained, relatively few positions need to be changed (SAALFELD, 1960). For this reason little activating energy is needed, and moreover the phase reaction may take place topotactically. Kinetics favour boehmite very much more than diaspore. Depending on the concentration-ratio Fe/Al, both Al OOH modifications may be formed simultaneously.

Ti-minerals

Several other elements dissolve and precipitate under similar conditions to aluminium. *Titanium* is an example.

KESSMANN (1966) worked with an experimental preparation of the TiO_2–Na_2O system. Low temperatures and small amounts of alkalis seem to favour stability of anatase rather than brookite (Fig.25).

The concentration of TiO_2 in laterites and bauxites ranges from 0.5–33 % (KATSURA et al., 1962). In the case of igneous source minerals it is mainly tied to ilmenite and titanomagnetite. In silicate structures such as augite, hornblende and biotite where Al^{3+} usually has the coordination number 6, it may be replaced by

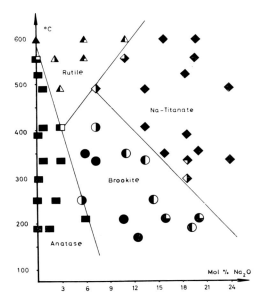

Fig.25. Equilibrium in the system TiO_2–Na_2O related to temperature. (After KESSMANN, 1966.)

Ti^{4+}. During weathering of titanomagnetite and ilmenite which are the most common sources of titanium, Fe and Ti migrate separately.

MATSUSAKA et al. (1965) discovered titanomagnetite with 21–25 mole % of TiO_2 to be a primary mineral in latosols on basalts of Hawaii.

Oxidation of Fe^{2+} initiates the formation of K-titanomaghemite (K = katmorphic).

Some of the titanomaghemite in the strongest weathered soils may have been formed by dehydration of lepidocrocite in the presence of Ti^{4+}. This has been named A-titanomaghemite (A = anamorphic).

In acidic and reducing environment titanium dissolves and is transported as Ti^{4+}. There is reprecipitation of titanium and iron as anatase and maghemite or hematite, respectively.

Whether titanium precipitates with iron or aluminium is a function of the Fe values. Experience shows that titanium enriches with aluminium in most laterites, indicating transport of Fe^{2+} and precipitation of $Fe(OH)_2$.

So far observations have shown titanium neomineralization in laterites to be confined almost entirely to anatase. The volume of anatase neomineralization depends on the quantities of titanium available. Up to 15% anatase is recorded from bauxites of the Bihar Mountains in India. It is not visible under the microscope, indicating fine dispersion throughout the bauxite.

The amount of titanium may even increase with extensive lateral solution transport. In weathering profiles of the Hawaiian Islands, SHERMAN (1952) observed titanium dioxide concretions up to 1 cm in diameter. These occur especially in the Fe-enriched A-horizon of the humic ferruginous soils.

Common epigenetic minerals

As reducing conditions prevail only in localized parts of bauxites, Fe^{2+} minerals are subordinate. The formation of different Fe^{2+} minerals depends not only on P_{O_2} but also on P_{CO_2} and P_{H_2S} (Fig.26).

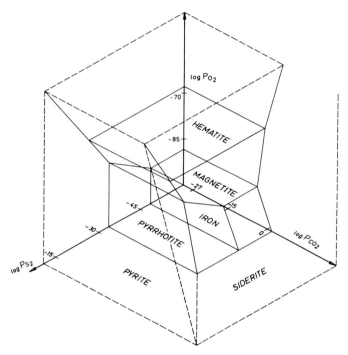

Fig.26. Stability relationship of some iron compounds as functions of P_{O_2}, P_{CO_2} and P_{S_2} at 25 °C and 1 atm., or greater total pressure. (After GARRELS and CHRIST, 1965.)

Bauxites may be supplied with S and CO_2 from polygenetic soil formation or overburden with lignites or peat bogs. Postdiagenetic minerals formed under fluctuating pH and Eh values are sulphides, sulphates and carbonates such as pyrite, marcasite, green vitriol, alunite, jarosite, gypsum, siderite and calcite. There is metasomatism or precipitation in joints and cavities.

In bauxites of the Urals (BENESLAVSKY, 1963) the following zeolites occur: chabasite, phillipsite, ferrierite, heulandite, and harmotome.

Chapter 4

CLASSIFICATION OF BAUXITES

SOILS AND WEATHERED ROCKS

Classification

Enrichment of aluminium may take place in various tropical and subtropical soil types on different source rocks. The most important soils are *latosols* and *andosols*, respectively, which may differ in character depending on local conditions.

Latosols

There is a comprehensive literature on latosols, and the reader is referred to MOHR (1938), KELLOGG (1949), PRESCOTT and PENDLETON (1952), SIVARAJA-SINGHMAM et al. (1962), DUCHAUFOUR (1965), DELVIGNE (1965), AUBERT (1965) and SÉGALEN (1965).

Several classifications of in situ latosols have been suggested in the past, based on specific conditions in various countries. MAIGNIEN (1964) discusses in detail distinct criteria for classification, and the co-existing classifications in the U.S.S.R., France, Portugal, Great Britain, Australia, the U.S.A., Belgium and the I.P.S.- and F.A.O.-systems.

The nomenclature established by the U.S. SOIL SURVEY STAFF (1960, pp. 238–244) distinguishes 10 soil orders and defines lateritic soils = latosols = oxisols = ox, as the 9th order. Their occurrence is commonly confined to old peneplains of tropical or subtropical regions.

So little is known about latosols that subdivisions are provisional. The subdivisions are based on characteristics of the upper 125 cm of the weathering profiles which is inadmissible in soil-genetic studies. The criteria referred to are: degree of lithification, base exchange capacity, and periodic phases of drying up and water impregnation.

In accordance with the French system of classification, to which other European systems are related, DUCHAUFOUR (1965) subdivides latosols (soils rich in sesquioides) into: (*1*) sols ferrugineux; (*2*) sols ferrallitiques: (*a*) sols rouges mediterranéens; (*b*) sols ferrugineux tropicaux; (*c*) sols ferrallitiques. This classification is adopted by the present author in principle.

Both the Mediterranean soils and the ferrallitic soils may become strongly enriched with aluminium.

PLATE I

1

2

3

4

5

Andosols

The Congress on the Correlation of Soils from Volcanic Ash in Tokyo in 1964 defined andosols as mineral soils derived from volcanic ash. They contain at least 50% amorphous substance and are commonly high in Al with gibbsite being the aluminium mineral. They have a high sorptive capacity, a relatively thick, friable dark A-horizon with substantial amounts of organic matter, a low bulk-density and a low stickiness. These soils form under humid and subhumid conditions.

In places they are an important source material for both autochthonous and reworked bauxites.

Catena

Co-existing soil types contemporaneously developed under different conditions are termed "catena" by the soil scientist or "facies" by the geologist.

The factors which may govern Al enrichment in certain catenas are: source rock, relief, climate and vegetation, ground water circulation.

The formation of aluminous latosols is favoured by:

(*a*) Aluminous source rocks with little quartz, such as alkali syenites and basic igneous rocks.

(*b*) Plateaus with cuestas or valley slopes.

(*c*) Periodical changes of humidity in tropical climates. A scheme of the relationship between soil type and climatic factors proposed by SÉGALEN (1965) is given in Fig.27. Very little is known about the influence of vegetation on Al enrichment.

(*d*) Intensive ground water circulation with the rapid removal of bases and silica.

Polygenetic soil formation and fossil soils

The weathering process and soil formation may progress rapidly under constant

---

PLATE I
Plateau type of bauxite on basalts.
   Macrofabrics:
1. Macrofabric of bauxite showing basaltic relic texture with columnar jointing; Mewasa I, Gujerat, India.
2. Vesicular texture of the upper iron crust. There is development of a dark hematitic skeleton intergrown with a light-coloured kaolinitic or gibbsitic matrix; Udagiri OE 1/1, India.
3. Vesicular texture of the upper iron crust; the soft kaolinitic part has been washed out; Manduapat M3/5a, India.
4. Partly destroyed upper iron crust; Manduapat M6/5, India.
5. Iron oxide granules and pebbles as remnants of a completely destroyed iron crust; Manduapat M6/6, India.

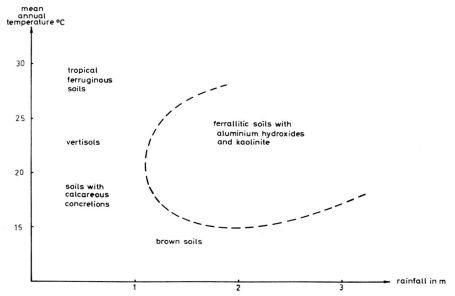

Fig.27. Relation of annual mean temperature and rainfall to soil formation. (After
SÉGALEN, 1965.)

climatic and ground water conditions on favourable source rocks with a high
percentage of readily weathering silicates. Soils formed in such manner are termed
"monogenetic soils". High-level areas commonly have long terrestrial periods.
Changes in climate and vegetation or tectonic movements or relief cause the soils
to adjust to the new environment. They transform into "polygenetic soils" or
are reworked and redeposited.

Many laterites, bauxites and terra rossa formations previously believed to
be Recent, are now known to be fossil. The laterites were degraded and younger
soils superimposed on them resulting in polygenetic soils. "Genuine" fossil latosols
are preserved only where rapidly covered by transgressive marine sediments,
preventing progressive soil transformation.

Soil sediments

If a period of intensive soil formation is succeeded by uplift, intensified relief
energy becomes effective and initiates downhill sliding or denudation of the soil.
Such sliding is common in high-level sections of present-day tropical areas. The
beginning of such soil movements has not yet been accurately dated but probably
occurred in Late Tertiary to Pleistocene time. In this book the term soil sediments
is applied to those soils which develop progressively while sliding downhill or
became reworked and redeposited.

Rhythmic sedimentation with intercalated fossil soils

If there is complete redeposition of the soils, soil formation processes, denudation and redeposition are the successive phases in terrestrial areas. For this reason the following series may develop at the same time high-level areas and basins, respectively:

high-level areas	*basins*
I. soil formation	
polygenetic transformation	↑ polygenetic transformation
mechanical reworking	soil formation
denudation	sedimentation
II. soil formation	polygenetic transformation
↓ etc.	soil formation
	sedimentation

ERHART (1956) applies the term "Biostasie" to periods of soil formation and "Rhexistasie" to phases of denudation. In periods of biostasis there is normal vegetation while phases of rhexistasis are characterized by the dying out or lack of vegetation due to soil erosion resulting from climatic changes or tectonic displacement.

The disappearance of vegetation initiates degradation and denudation of soils and weathering zones. The period of rhexistasis is characterized by mechanical reworking, whereas the phase of biostasis is characterized by chemical decomposition.

Erhart distinguishes five types of fossil soils: (*a*) autochthonous soil protected by sediments; (*b*) autochthonous relic soils without protective cover; (*c*) allochthonous soils or pseudo-fossil soils (on slopes); (*d*) mixed soils or heterosols; (*e*) pedolites or allochthonous soils altered by diagenesis.

In all these soils Al may be enriched to such an extent as to become economically important. Preservation of the various types depends largely on geotectonics. On shields with long terrestrial periods latosols may outlast very long geological epochs (latosols in the highlands of northern Ethiopia are of Jurassic age at least) or they degrade or erode as a result of climatic changes (Sudan). Fossil bauxite deposits derived from such latosols are widespread on the earth's crust with undisturbed profiles where covered by protective sediments (Arkansas, Gujerat).

Relationship between structural development and bauxite formation

Bauxite deposits interbedded with sediments and preserved by rapid transgression

PLATE II

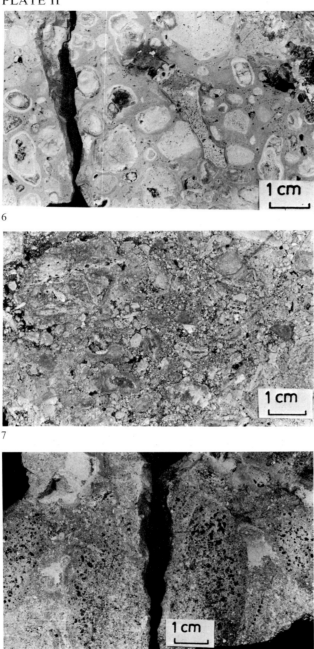

6

7

8

are part of the stable shelf (Surinam, Weipa) or unstable shelf regions (Mediterranean areas).

In Russian literature bauxites are classified into plateau bauxites and "geosynclinal" bauxites. Plateau bauxites are deposits on ancient shields and on young eruptiva, whilst "geosynclinal" bauxites—mostly in the unstable shelf region or on welts made of carbonate facies—are interbedded with clastic or carbonate sediments. Bauxites are not known to have formed in true geosynclines. The formation and preservation of bauxite deposits is governed by specific relationships between structural environment and time.

Historical record of bauxite formations

The oldest larger bauxite deposits so far observed occur in Early Palaeozoic sediments. STRACHOW (1961) tried to compile the known deposits in chronological order. He found that bauxite was often associated with coal or sedimentary iron ores of neighbouring areas. He reconstructed climatic zones which moved from north to south during the earth's history. In many cases accurate dating is not possible because of significant unconformities between both bauxite and the overlying and bauxite and the underlying strata. For this reason statements on the age of outcropping bauxites differ widely. However, there are certain periods in the earth's history which favoured bauxite formation.

TABLE of periods of bauxite formation (modified after PATTERSON, 1967)

Cenozoic

Recent and Pleistocene:	tropical lateritic bauxite of *Panama, Costa Rica, Hawaii, Fiji, British Mandated Solomon Islands*
Tertiary (undifferentiated)	gibbsitic bauxite deposits in *Brazil, Venezuela, Guinea, Ivory Coast, Ghana,* and *Western Australia* are at the surface but overlie older rocks and may have formed earlier than in Pleistocene time

PLATE II

Bauxite on basalts

6. Bauxite with vesicular texture. Note the skeleton of spongy gibbsitic bauxite in a dense gibbsitic matrix; Dudhi Pahar D2, India.

7. Gibbsitic bauxite with breccia-like texture: coarse- and fine-grained angular fragments of spongy bauxite in a dark matrix rich in iron; Bagru Hill V/1/10, India.

8. Gibbsitic bauxite with spongy texture; Bagru Hill V/1/11, India.

Miocene

U.S.A. (Oregon): lateritic gibbsitic bauxite
Germany (Vogelsberg Mountains): ferruginous gibbsitic bauxite
Australia (Victoria): gibbsitic bauxite
Jamaica, Haiti, Dominican Republic: gibbsite, capping Eocene and Oligocene rocks and probably formed intermittently in Miocene and post-Miocene times

Oligocene

Northern Ireland (County Antrim): ferruginous gibbsitic bauxite with overlying and underlying basalt flows

Eocene

Guyana, Surinam, French Guiana: gibbsite
Australia (Queensland, Northern Territory, Tasmania): gibbsitic bauxite, with underlying Cretaceous, Jurassic and Precambrian rocks
U.S.A. (Arkansas, Alabama, Georgia): gibbsite, with underlying Paleocene (Midway group) and older rocks, and overlying mainly Early Eocene rocks (Wilcox group)
India (Deccan peninsula): gibbsitic bauxite and laterite, developed on Deccan trap and Precambrian charnockite
India (Kashmir-Jammu): diaspore, with underlying Jurassic and overlying Eocene rocks
Italy: gibbsite and boehmite
Yugoslavia: boehmite and gibbsite, with underlying Early Eocene and Late Cretaceous rocks

Mesozoic
Cretaceous
Upper
Turonian

Greece: boehmite (and diaspore), with underlying Early Cretaceous rocks and overlying Late Cretaceous rocks
Yugoslavia: boehmite
Austria: boehmite
Turkey: boehmite
Hungary: boehmite chiefly

Lower	
Albian-Aptian	*Spain, France (Var, Hérault), Italy, Greece, Turkey*: boehmite chiefly (and diaspore)
Barremian	*France (Ariège)*: boehmite and diaspore
	Hungary: boehmite chiefly
Neocomian	*Hungary, Yugoslavia, Rumania (Bihar)*: boehmite chiefly (and diaspore)
Jurassic	
Triassic	
Upper	*Yugoslavia (Croatia)*: boehmite, with underlying Middle Triassic rocks
Lower	*Mainland China (Poshan)*: diaspore, with underlying Permian and Carboniferous rocks
Palaeozoic:	
Permian	*Turkey*: diaspore
Pennsylvanian	*U.S.A.*: diaspore in *Missouri* and *Pennsylvania* interbedded with rocks of Pottsville age
Mississippian	*Mainland China (Yunnan)*: boehmite, with underlying Devonian rocks
	U.S.S.R. (Tikhvin): bauxite, with underlying Devonian rocks
	Mainland China (Kweichow): diaspore, with underlying Ordovician rocks
Devonian	*U.S.S.R. (the Urals)*: bauxite, in Middle and Early Devonian rocks
Precambrian:	
Late Proterozoic	*U.S.S.R. (Bokson, Siberia)*: boehmite and diaspore (?)

The record demonstrates that there were distinct main phases of bauxite formation in the earth's history. The big deposits formed during Eocene times occur worldwide. The Cretaceous deposits, the majority of which are of Early Cretaceous age, have a specific regional distribution pattern, too. In Palaeozoic times the Mississippian and Pennsylvanian periods were characterized by widespread bauxite formation.

A possibly greater importance with respect to bauxite formation must be ascribed to Lower Devonian time as evidenced by the few bauxite formations preserved.

It is important to realize that present-day climatic conditions do not favour latosol formation. Nowadays there are only a few small localities with the extreme climatic conditions that cause ferrallite formation. Modern climates do not

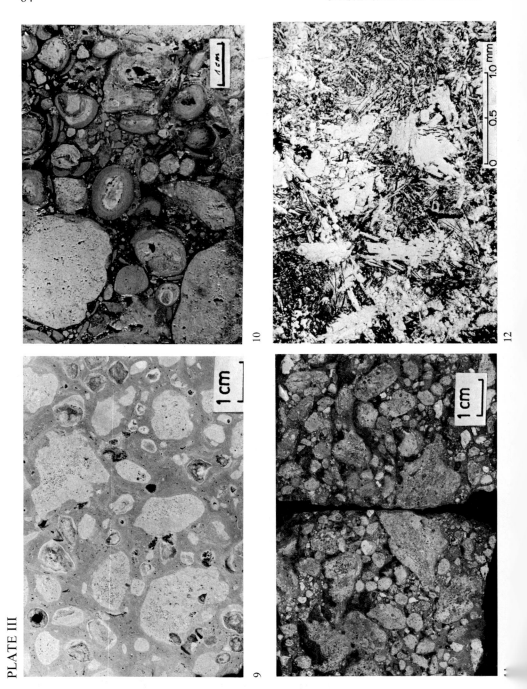

PLATE III

produce regional gibbsite or hematite formation; goethite is the dominant recent iron mineral. Apparently there were periods of extreme climatic conditions in the earth's history which caused ferrallites to be formed worldwide. These periods were characterized by thick terrestrial or paralic strata on shields or in shelf regions, such as coal, uniform sandstones and pure kaolinite clays.

ORE DEPOSITS

Bauxites are preferably incorporated in an ore-deposit and petrological classification rather than in a soil classification. The criteria suggested for classification are: (*a*) autochthonous or allochthonous origin; (*b*) chemical composition; (*c*) mineralogy; (*d*) textures.

Autochthonous or allochthonous origin

Bauxites may be grouped as follows:

(*1*) autochthonous bauxites on fresh igneous, metamorphic and sedimentary rocks: (*a*) primary formations; (*b*) fossil, polygenetic, altered bauxites;

(*2*) allochthonous bauxite sediments: (*a*) coarse clastic bauxite sediments; (*b*) fine clastic-colloidal bauxite sediments.

Chemical composition

Definition of the terms: allite, siallite, ferrallite

The terms allite, siallite and ferrallite were introduced by HARRASSOWITZ (1926) to denote a predominance of Al, Si + Al and Al + Fe, respectively. For classification of allite and siallite Harrassowitz used the molecular ratio Ki = SiO_2/Al_2O_3 under the assumption that there is only combined silica. The classification established by him distinguishes between:

PLATE III

Bauxite on basalts
9. The preliminary stage of pisolitic texture in bauxite: angular fragments of gibbsitic spongy bauxite with initial incrustations in dense gibbsitic matrix; Dudhi Pahar D2/6, India.
10. Pisolitic bauxite with nuclei made of gibbsitic spongy bauxite and shells consisting of boehmitic bauxite. The shells often break off and become cemented by a dense matrix rich in iron; Dudhi Pahar, India.
11. Breccia-like texture resulting from leaching of saprolite. The light angular fragments of spongy gibbsitic bauxite rich in kaolinite rest in a dark ferruginous kaolinitic matrix; Udagiri BE 4/9, India.
    Microfabrics:
12. Relic texture with pseudomorphism of gibbsite after feldspar basaltic of rock; Bagru Hill, India.

PLATE IV

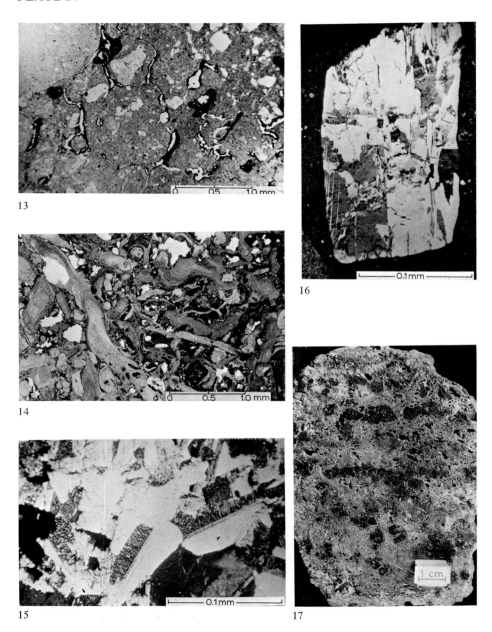

13

14

15

16

17

Ki > 2.0   = kaolinite clays
Ki = 2.0–1.3   = siallites
Ki < 1.3   = allites

PEDRO (1966) showed that the classification of Harrassowitz was obsolete for two reasons: (*1*) Silicate layers, the index of which is greater or smaller than Ki = 2, participate in lateritic weathering. The molecular ratio Ki is < 1.3 in silicate rocks made of amesite, sudoite or margarite (coordination number IV and VI for Al), requiring classification as allites. (*2*) Silicates with Ki > 2.0, such as orthoclase, albite, pyrophyllite and anauxite (Al with coordination number VI only) might be enriched with considerable amounts of hydrated Al oxides without being referred to as allites.

There are bauxites in Yugoslavia and Siberia with Ki ~ 1.3 that consist of aluminium hydroxides and pyrophyllite in approximately equal proportions, as described by BARDOSSY (1965).

The decisive criterion for classification is the amount of free aluminium. For this reason numerical values of Ki are meaningless without known mineral composition. The parameter Ki = 2 is only pertinent in the characterization of lateritic material if it consists of mixtures of well-crystallized kaolinite/gibbsite or kaolinite/boehmite.

There are several proposals that weathered rocks be classified on the basis of SiO_2/Al_2O_3 or combined Al_2O_3/free Al_2O_3 ratios (PEDRO, 1966). The following are noteworthy:

HARRASSOWITZ (1926); DE WEISSE (1948); AUBERT (1954): Ki = SiO_2/Al_2O_3 mole %.

DE LAPPARENT (1930): A = Al_2O_3/SiO_2 mole %.

GORDON et al. (1958): x = $Al_2O_3/1.1\ SiO_2$ mole %.

PEDRO (1966) and BARDOSSY (1963): combined Al_2O_3/free Al_2O_3.

The most suitable classification (Fig.28) appears to have been established by BARDOSSY (1963) and PEDRO (1966). However, their classification does not conform to ore-deposit classifications since bauxites with > 10% kaolinite are generally

---

PLATE IV
Bauxite on basalts
13. Bauxite with gel-like texture also showing initial development of polygonal elongated pore space; Mewasa I, India.
14. Vermicular texture in bauxite originating from digging fauna; Udagiri OE 1/6, India.
15. Recrystallized matrix with large gibbsite twins; Udagiri OE 1/9, India.
16. Twinning of gibbsite crystal; Mewasa I/6, India.

Bauxite on charnockite
17. Spongy bauxite texture; goethite pseudomorph after garnet (dark spots); Yercaud II/2, India.

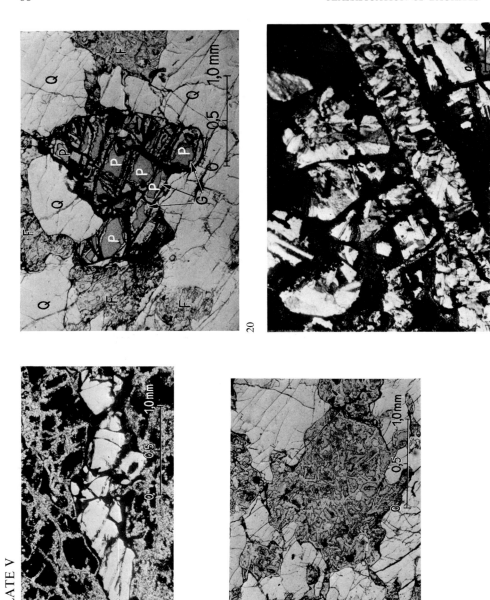

PLATE V

$Ki = \dfrac{SiO_2}{Al_2O_3}$	0 0,1 0,5 0,6 0,7 1 1,3 1,55 1,6 1,7 2
free Al_2O_3 (%)	100 95 90 80 70 60 50 40 35 30 20 10 0

	free Al_2O_3 regions				
AUBERT (1954) BOTHELO de COSTA (1959)	Sols leviferrallitiques			Sols ferrallitiques typiques	Sols faiblement ferrallitiques
HARRASSOWITZ (1926)	allites			siallites	
LACROIX (1913)	Laté-rites	Latérite silicatée	Argile latéritique		Argile
de WEISSE (1948)	Bauxite		Terra rossa		
GORDON et al. (1958)	bauxite		bauxitic clay		kaolinitic clay
BARDOSSY (1963)	bauxite		clayey bauxite		bauxitic clay
PEDRO (1966)	bauxite		bauxite kaolinitique	argile bauxitique	argile
VALETON (1968)	high-quality bauxite	low-quality bauxite	kaolinitic bauxite	bauxitic clay	clay

Fig.28. Aluminous rocks derived from weathering. A classification by various authors based on Ki-index or free alumina content.

uneconomical. For this reason bauxites require further subdivision into high-quality bauxites and low-quality bauxites with 0–10% combined Al_2O_3 and 10–25% combined Al_2O_3, respectively.

Chemical balance

Inquiries into the chemical balance of lateritic weathering are supplementary to questions of classification based on the content of free and combined aluminium.

MILLOT and BONIFAS (1955) quantitatively calculated the chemical balance of the mineral transformations in laterites and bauxites with constant volume proven by excellent preservation of relic textures of the source rock. This iso-

---

PLATE V
Bauxite on charnockite
18. Relics of quartz "fractured" by chemical dissolution surrounded by a gibbsitic skeleton. Dark areas are pore space (crossed nicols); Yercaud II/1, India.
19. Gibbsitic pseudomorphism after feldspar with "fossil" feldspar cleavage. There are relics of strongly corroded quartz grains. Gibbsite recrystallized in pore space at a later stage; Yercaud II/1, India.
20. Initial replacement of pyroxene by goethite along cleavage followed by crystallization of gibbsite ± (G) laths in pore space ± (P). Surrounding feldspar (grey) and quartz (white); Yercaud II/1, India.
21. Pyroxene replaced by highly porous goethite skeleton with coarse-grained gibbsite filling former pore spaces (crossed nicols); Yercaud I/4, India.

PLATE VI

TABLE XII

WEATHERING OF SYENITE OF THE ISLAND LOS (after BONIFAS, 1959)

	Fresh syenite (I)	Weathered syenite (II)	Porous weathering product (III)	Differences in weight between I ánd III	Variation from composition of the syenite source rock (%)
Weight (g/cm³)	258	219	154		
SiO₂	148.0	127	4.8	−143.2	− 96
Al₂O₃	45.0	47.5	86.2	+ 41.2	+ 91
Fe	11.9	5.3	8.1	− 3.8	− 32
CaO	6.7	0.8	0.3	− 6.4	− 96
MgO	2.8	0.8	—	− 2.8	− 100
Na₂O	15.5	8.8	n.d.	n.d.	
K₂O	18.1	17.7	n.d.	n.d.	
TiO₂	1.8	1.3	2.6	+ 0.8	+ 44
MnO₂	1.0	0.2	0.1	− 0.9	− 90
H₂O	3.3	6.8	45.7	+ 42.4	−1,280

volumetric method is as yet the only means which provides absolute data on the movement of chemical elements in defined directions. The principle of this calculation is to compare mole % of elements in 1 cm³ of fresh rock and corresponding weathered rock (Table XII).

If relic textures are not observable, calculations applicable are listed below.

KÖSTER (1961) referred to the index method of MARSHALL (1940, 1942) which is based on using intact minerals as the reference for transformations in individual horizons. He uses quartz as the index mineral. For bauxites, however, quartz is not a suitable mineral as it may be dissolved (Weipa/Queensland). There is only very little zircon in bauxites, which is likely to be unstable as demonstrated by secondary ZrO_2 precipitations. Chromite and ilmenite are not always sufficiently stable either.

STRENG (1858, 1860) introduced the oxide of specific elements such as TiO_2 or Al_2O_3 as the invariable factor = 100% in all horizons of the weathering profile.

PLATE VI

Bauxite on sedimentary rocks.

22. Bauxite formed by replacement of stratified arkosic Eocene sediments; Onverdacht, Surinam.
23. Karst relief of Middle Jurassic carbonate bedrock; Combecave, Var, France.
24. Complete profile of Early Cretaceous boehmite bauxite with relics of parts rich in primary hematite (dark areas). There was extensive leaching of hematite during late stages of diagenesis, leaving behind a highly aluminous bauxite body that is kaolinitic in the upper dense part; Mazaugues, France.

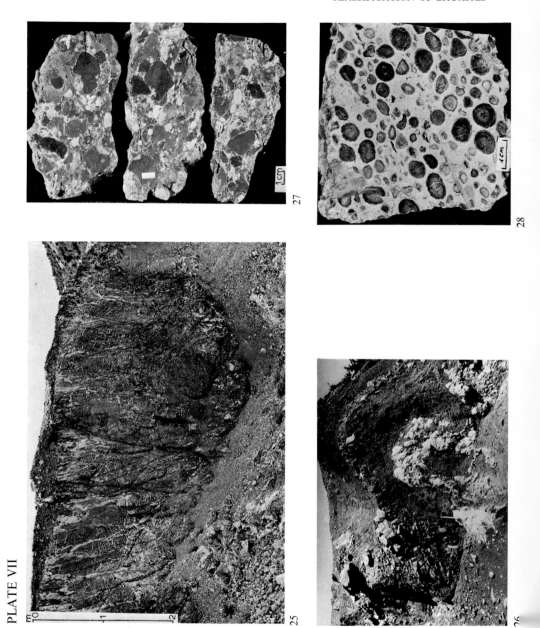

PLATE VII

The method is not applicable to bauxites as there are no standard elements with constant proportions in weathering profiles.

In order to decide on enrichment or solution in bauxites without relic textures, the reader is referred to the method of GROSSER (1932a, b; 1935) which is based on division of the weight percentages of the individual oxides in weathering profiles by the corresponding values of fresh rock. It is thus possible to calculate enrichment factors for specific elements of the weathering profile. However, the method as used for the bauxites of Arkansas does not provide the solution for absolute or relative enrichment of specific components.

Mineralogical composition

The most practical classification is based on mineral groups as proposed by KONTA (1958) because both clastic components (quartz, ore minerals, etc.) and neomineralization may be characteristic features of the bauxites. BARDOSSY (1963) widened the scheme and included the clastic fraction. The mineral groups defined by him are:

(*1*) allitic minerals: gibbsite, nordstrandite, boehmite, diaspore, corundum; (*2*) ferritic minerals: goethite, lepidocrocite, hematite, magnetite, maghemite; (*3*) clay minerals: predominantly kaolinite group minerals, associated with illite-montmorillonite–chlorite group minerals; (*4*) clastic minerals: quartz, heavy minerals, ore minerals.

Fig.29 shows a classification scheme proposed by the present author, modified from BARDOSSY (1963).

The process chosen for alumina production depends on the natural mineral assemblage of the ore deposit. For this reason accurate maps recording the regional distribution of these minerals in bauxites are required for industrial exploitation of the deposits. The geological maps of Greece, for instance, record "soluble" boehmite and "insoluble" diaspore bauxites with respect to the Bayer process.

PLATE VII
Bauxite on sedimentary rocks
25. Face of an Early Cretaceous bauxite body near Combecave, France, demonstrating post-diagenetic removal of iron starting from the top of the ore.
26. Postdiagenetic degradation of Late Cretaceous bauxite caused by postbauxitic karstification of the limestone bedrock and formation of terra rossa from bauxite; Greece.
27. Clastic breccia-like bauxite of Early Cretaceous age with pebbles of iron crust, kaolinitic clay and red clay sedimented in close proximity to the lateritic source rock of the Maure-Esterelle Massif; Pas de Recou, Var, France.
28 Early Cretaceous pisolitic diaspore bauxite with elongation and preferred orientation of the pisolites (flow texture). Dehydration cracks of pisolites are filled with kaolinite; Péreille, Ariège, France.

PLATE VIII

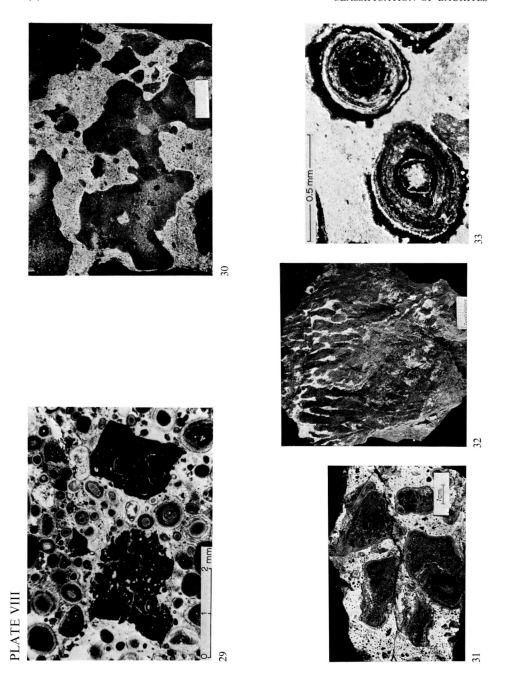

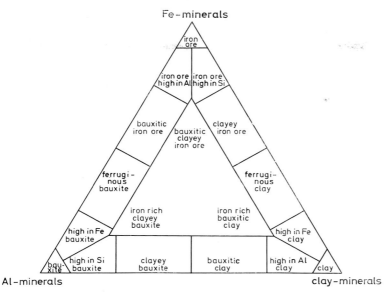

Fig.29. Proposed classification of bauxites and aluminous iron ores and clays. (Modified from BARDOSSY, 1963.)

Fabrics

The discussion of textures is based on two distinct types of source rocks:

(*1*) Unlithified rocks as red earths, red loams, and brown loams. Both relic textures and textures of neomineralization are difficult to find in these rocks.

(*2*) Solid rocks, the textures of which have withstood weathering. The textures may be subdivided into *relic* textures of the source rocks and textures formed during *neomineralization*.

PLATE VIII

Bauxite on sedimentary rocks

29. There was extensive leaching of iron during the late stages of diagenesis or in postdiagenetic time, leaving behind fringed remnants of originally widely dispersed hematite. The crowded pisolites formed during early stages of diagenesis are clouded by the "relic concretions" but are easily seen in the bleached areas; Late Cretaceous, Greece.

30. Incomplete deferrification of Late Cretaceous bauxite in Greece during late stages of diagenesis. Note the dark angular remnants of hematite.

31. Hematite relics (dark) with secondary goethite crust (grey) and bleached matrix. Hematite formed during early stages of diagenesis, while goethite is of late diagenetic or postdiagenetic origin; Mazaugues, Var, France.

32. Manganiferous seam between bauxite and bedrock approximately 10 cm thick showing vesicular texture: white = kaolinite, dark = todorokite; Early Cretaceous, La Rouge, Var, France.

33. Pisolites showing alternating zones of hematite and boehmite embedded in fine-grained boehmitic matrix. The outer shells, rich in iron, are partly replaced by matrix; boehmite facies; Mazaugues, Var, France.

PLATE IX

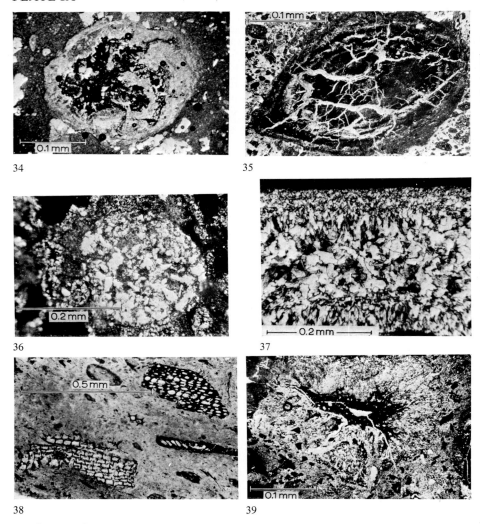

34

35

36

37

38

39

Bauxite on sedimentary rocks

34. Concretion of siderite, hematite and magnetite in the centre fringed by kaolinite, embedded in a diasporic matrix. Diaspore facies; Early Cretaceous, Péreille, Ariège, France.

35. Diaspore-rich oolite still in its plastic stage deformed during diagenesis with shrinkage cracks filled by kaolinite. Diaspore facies; Early Cretaceous, Péreille, Ariège, France.

36. Oolite showing large diaspore crystals lacking orientation. Diaspore facies; Early Cretaceous, Aude, France.

37. Fissure filling of coarse-grained diaspore by precipitation from solution; Late Cretaceous, Greece.

38. Fossilized plant relics with a coaly skeleton in fossil root horizon embedded in fine-grained boehmitic matrix; topmost part of the profile; Early Cretaceous, Péreille, Ariège, France.

39. Concentration of kaolinite (white) and hematite (dark) demonstrating former capillary systems (probably a former root); topmost part of the profile; Early Cretaceous, Mazaugues, Var, France.

Long ago NEWBOLD (1844) pointed to relic textures of source rocks commonly preserved in both macro- and microdimensions and concluded that in situ transformation had occurred. Relic textures not only characterize the source rock, but also provide answers as to how much certain components became enriched.

Textures of neomineralization are superimposed on relic textures. In spite of several attempts, no detailed classification has as yet been established for both micro- and macrotextures formed during neomineralization. With respect to macrotextures, Al-rich latosols are only distinguished as having no or marked zoning (ground water laterites). The ground water laterites have been classified in various ways as shown in Table XIII.

The B-horizon and the lower part of the BL-horizon are the main zones of aluminium enrichment. The fresh rocks are porous and plastic, and may be cut with a knife. The profiles are commonly 10–30 m thick. The textures of neomineralization within individual horizons of autochthonous bauxite may be classified as follows:

Skeleton bauxite. These bauxites are characterized by skeleton-type preservation of micro- and macrofabrics of the source rocks caused by relative enrichment of less soluble elements (Plate I, 1; III, 12; IV, 17; V, 18–21). The relic textures derived from igneous rocks may be granular or granitic.

Gel-like textures. Absolute enrichment of Al or Fe resulted in destruction of relic textures in microdimensions followed by rearrangement in spherical or shelly precipitation fabrics (Plate IV, 13).

TABLE XIII

CLASSIFICATIONS OF GROUND WATER LATERITES

NEWBOLD (1844)	LACROIX (1913-14)	WALTHER (1915)	MOHR (1938)	PRESCOTT and PENDLETON (1952)	GORDON et al. (1958)	*Soil horizons*
				A1	A1	A1
				A2	A2	A2
laterite crust (murrum)	cuirasse de fer, zone de concrétion	Eisenkruste	ironstone crust	laterite	zone of concretion	BL
reddish and yellow dust		Fleckenzone	mottled zone	mottled zone	zone of leaching	B
lithomarge	zone de départ	Bleichzone	pallid zone (zone of cementation)	pallid zone	under clay (saprolite)	B/C
parent rock	roche mère	Ausgangs-gestein	parent rock	parent rock	parent rock	C

Two types of textures may be distinguished: (*a*) dense textures with negligible pore space; (*b*) spongy textures with high porosity. The pores are distributed irregularly throughout the rock, both in micro- and macrodimensions (Plate II, 8).

Breccia-like texture. Intensive solution may cause the primary textures to collapse. The observer notes fragments of relic textures cemented by various matrices in their respective local environment.

Oolitic, pisolitic and concretionary textures. Rhythmic spherical precipitation on cores results in: (*a*) oolitic textures (< 2 mm in diameter); (*b*) pisolitic textures (2–20 mm in diameter); (*c*) concretionary textures (>20 mm in diameter). The textures are linked with spongy textures by transitional zones. Rhythmic dehydration and rehydration causes the spongy textures to transform into breccia-like textures. There are angular particles (several mm to several cm in diameter) with a spongy texture that rest in a matrix of dense texture (Plate III, 9, 10).

Vesicular textures. In cm or dm dimensions, intimately interwoven vein or tubular texture systems develop if two phases such as allitic or siallitic and ferritic phases separate from gel (Plate I, 2; II, 6). They must be distinguished clearly from vermicular textures.

Vermicular textures originate from digging fauna (Plate IV, 14).

Cellular bauxites. The textural term refers to bauxites characterized by cavities caused by selective solution in macrodimensions.

Earthy bauxite. This is a soft unlithified bauxite of sedimentary origin.

Nodular texture. Bauxites with earthy texture may contain small concretions of variable mineralogical composition, but are chiefly made of gibbsite, and dispersed throughout the rock.

Textures with relic concretions. These textures develop in ferrallitic bauxites from the epigenetic selective solution of iron in large parts of the bauxite. Proportions of the bauxite not affected by solution of iron are referred to as relic concretions (Plate VIII, 29, 30). The higher density of the relic concretions cause the precipitation of secondary coatings (Plate VIII, 31).

Impregnation concretions. This type of concretion formed during the final stage of diagenesis or even during epigenesis and originates from mobilization of iron and precipitation of iron minerals in distinct areas dispersed throughout the matrix.

Chapter 5

DESCRIPTION OF TYPES OF DEPOSITS

BAUXITES ON IGNEOUS AND METAMORPHIC ROCKS

Bauxite deposits of this type are known to have formed throughout the earth's history on plateaus during long terrestrial periods. In most cases they eroded, but bauxitic sediments which surround the fossil plateaus prove their original existence. Intercalated diaspore-boehmite bauxites in Cambrian limestones in the eastern part of the Sayan Mountains in the general area south of Lake Baikal (U.S.S.R.) are derived from the oldest known lateritic bauxites on igneous or metamorphic rocks. In the Onega district (U.S.S.R.) laterite bauxites of Late Devonian to Early Carboniferous age occur on basalts. Also well known are Jurassic to Cretaceous laterite bauxites on ophiolites of the sub-Pellagonic zone in Greece and on ophiolites of similar age in Yugoslavia. Frequently links of weathered plateau and foreland covered with redeposited weathering products are disrupted through deep erosion. Therefore, complete reconstruction of areas of denudation and distribution is not always possible.

However, since Late Cretaceous and Early Tertiary times large parts of the world were similarly shaped and with similar climatic conditions as occur at present.

A belt of bauxite deposits, formed mainly during Late Cretaceous and Early Tertiary times, spreads over continent-wide peneplains on Precambrian rocks of South America, Africa, Southeast Asia and Australia. Accurate age determinations on peneplains of these areas are difficult or impossible to achieve. However, in Tertiary times bauxites frequently formed on littoral sediments which consist of reworked detrital laterites or in situ bauxites. Therefore, the age of the plateau laterites is at least Early Tertiary.

Also Early Tertiary (Eocene) in age are the bauxites on nepheline syenites of Arkansas, on the Deccan Plateau in India and on basalts in southeastern Australia. However, the laterite bauxites on basalts of Oregon (U.S.A.), Vogelsberg (Germany) and Antrim (Ireland) formed during Late Tertiary times. The youngest lavas capped by bauxites, which can be dated precisely, occur in Hawaii and are 10,000 years old. Valleys already cut into the bauxites and recent flows are not lateritized. Possibly the main phase of bauxite formation occurred there in Alleröd time.

The scope of present-day transformations may be studied best in areas with Recent volcanic tuff. Extensive research is undertaken on the so-called "andosols" in areas of the Circumpacific Province. However, our knowledge concerning the extent and speed of transformation is still incomplete.

In spite of the continuous movement of Fe, Al and Si to date, all observations from Tertiary laterite bauxites indicate not further development but destruction of laterite profiles. Movements of the earth's crust resulted in different morphology, ground water movements and vegetation, and the old laterites represent poly-genetically altered soils.

Two types of bauxite deposits are distinguished on the basis of facies (catena):

(*1*) The slope type (gibbsite type) which forms lens-shaped bodies on the slope. It does not develop a typical vertical profile. Gibbsite is the main Al mineral.

(*2*) The plateau type (gibbsite-boehmite type) which is a lateral facies of the plateau laterites along valley slopes with good drainage and marked profile differentiation.

Slope type

This type does not form large and important deposits. Because of its morphological exposure, it may be fossilized only if rapidly covered by younger sediments. The type which occurs widespread on basement rocks is known only from the Tertiary, Pleistocene and Recent. However, it is particularly important for the study of the transformations and paragenesis of bauxites.

Examples of slope types investigated, occur on: basic to intermediate volcanic rocks (Hawaii; Pleistocene to sub-Recent), acid metamorphic rocks (Ivory Coast; Late Tertiary to sub-Recent), and charnockite in southern India.

Hawaii

Young ferrallitization (probably Alleröd time) which led to substantial enrichment of aluminium hydroxides was first found on the Hawaiian Islands. Transport in solution in downward direction parallel to the slope in an area with extremely high rates of rainfall and good drainage, resulted in catenas with soil zones enriched with aluminium hydroxides. Old land surfaces do not exist, hence any impregnations with solutions from high plateau soils may be excluded.

The islands emerged from the Pacific only during Early Tertiary time and are composed of a series of olivine basalts, alkali basalts (nepheline and melilite basalts and basanites), trachytes and tuff. Basalt flows of the Kola series, Pleistocene in age (10,000 years), were transformed into bauxitic "latosol". There are no "latosols" on Recent lavas.

Morphology, climate, vegetation and ground water movements govern the formation of the soil type. If one assumes that these factors have remained constant over the past 10,000 years in the tropics, conditions governing bauxite formation may be read from present-day environment. The islands are cone-shaped with more or less steep slopes. The highest summit of the Kauai Island rises to 1,567 m. The river valleys average 30 m (max. 140 m) in depth. The Hawaiian Islands are situated in the belt of the northeasterly trade winds. Rainfall rates are extremely high on the northern and eastern slopes, and increase from the coast to the level of cloud cover at approximately 1,000 m, but decrease rapidly towards higher terrain. The following rainfall rates from Kauai, the island known to be the wettest, were recorded by ABBOT (1958):

precipitation	north coast	southwest coast	summit of Mt. Waialeate
mm/annum	2,250	500	11,240

Temperatures depend strongly on altitudes and corresponded to a subtropical climate at about 600 m.

The lava flows of the Kola series are always inclined (Fig.31). There are vertical textures in each flow. The welded flow breccia of glass and slag at the top of the basalt flow (klinkerzone) is permeable and enables the ground water to circulate both vertically and horizontally over long distances. The water table fluctuates in the dry and rainy seasons. Normally the ground water slowly percolates through the weathering zone until it reaches the boundary of saprolites (decomposed rock in B/C horizons) to fresh basalt, causing the solutions to flow laterally. The water table was analyzed (Table XIV) from several wells of the northern and eastern side of Kauai Island (PATTERSON and ROBERTSON, 1961).

The pH of the ground water ranges from 4.0 to 5.9 in wells in latosols (holes *1–6*), but is 7.6 to 7.8 in wells in fresh basalt (holes *7* and *8* in Fig.31).

The amount of HCO_3^- dissolved from latosols is negligible but increases to 78–96 p.p.m. from basalts. The total amount of salts in ground water solution is much less in soil (28–42 p.p.m.) than in fresh basalts (167–201 p.p.m.) from which silica, alkalis and the alkaline earths mainly dissolve. This is explained by both the relatively fast drainage in the weathering profile and the occurrence of soluble minerals of the fresh basalt.

Soil formations. The warm humid climate caused the formation of deeply weathered latosols. The classification given by SHERMAN (1958) according to location is presented in Table XV.

These latosols may be 15–30 m thick. The thickness and nature of the soils depend on the texture of the source rock, climatic and morphological exposure, and hence on intensity of ground water circulation. There is particularly fast weathering of the Kola series source rocks with no free quartz and a low silica content because of the original vitreous matrix with porous texture.

SHERMAN (1958) illustrated the relationship of increasing aluminium content

TABLE XIV

CHEMICAL ANALYSES[1] OF WATER SAMPLES FROM WEATHERED AND FRESH BASALTS OF THE KOLA VOLCANIC SERIES, KAUAI, HAWAII

Hole No.	Parts per million														
	SiO_2	Al	Fe	Mn	Cr	Ca	Mg	Na	K	Li	NH_4	HCO_3	CO_3	OH	SO_4
1	2.9	0.11	0.00	0.26	0.000	0.8	1.9	11	0.6	0.00	0.02	1	0	0	2.7
2	1.2	0.00	0.18	0.13	—	0.9	1.9	6.3	0.6	0.00	0.02	6	0	0	4.0
3	4.1	0.01	0.00	0.07	0.000	1.0	1.2	6.9	0.3	0.00	0.03	3	0	0	2.3
4	3.0	0.04	0.00	0.12	0.000	0.4	0.7	7.2	0.0	0.00	0.02	2	0	0	3.2
5	2.6	0.03	0.06	0.54	0.000	1.0	2.3	8.6	0.6	0.00	0.01	6	0	0	6.5
6	2.5	0.12	0.00	0.11	0.000	0.6	1.5	8.5	0.5	0.00	0.02	0	0	0	2.7
7	32	0.03	0.00	0.00	—	13	14	30	1.3	0.00	—	96	0	0	37
8	42	0.06	0.00	0.00	0.000	13	11	21	0.8	0.00	0.01	78	0	0	5.2

[1] Analysts A. S. Vandenburgh and C. E. Robertson, in PATTERSON and ROBERTSON (1961); samples are from weathered basalt, except those from holes 7 and 8, which are from fresh basalt.

TABLE XV

THE SOIL SERIES OF THE SOIL FAMILIES OF THE VARIOUS HAWAIIAN GREAT SOIL GROUPS WHICH HAVE MORE THAN 10 PERCENT FREE ALUMINA CONTENT

Great Soil Group	Soil family	
	name	location
Aluminous Ferruginous Latosol	Halii	eastern half of Kauai
Ferruginous Humic Latosol	Haiku	eastern half of Kauai
	Puhi	eastern half of Kauai
	Haiku	east Maui
	Naiwa	west Maui
Hydrol Humic Latosol	Akaka	Hawaii
	Hilo	Hawaii
	Honokaa	Hawaii
	Koolau	Kauai
		east Maui
Latosolic Brown Forest	Olinda	east Maui
Humic Latosol	Honolua	west Maui

Hole No.	Parts per million							Cation total	Anion total	Dissolved solids calculated (p.p.m.)	Hardness as CaCO₃ (p.p.m.)	N.C. hardness as CaCO₃ (p.p.m.)	Specific conductance (microhms at 25 °C)	pH
	Cl	F	I	NO₂	NO₃	PO₄	B							
1	19	0.1	0.008	0	0.6	0.02	0.04 0.70	0.64	41	10	9	90	4.6	
2	9.1	0.1	0.006	0	0.7	0.04	0.02 0.49	0.46	28	10	5	60	5.1	
3	12	0.1	0.003	0	0.4	0.03	0.03 0.46	0.46	30	7.4	5	56	5.6	
4	12	0.2	0.006	0	0.1	0.02	0.03 0.39	0.45	28	4.0	2	54	4.8	
5	16	0.1	0.004	0	0.9	0.02	0.05 0.63	0.71	42	12	7	78	5.3	
6	13	0.1	0.002	0	0.8	0.00	0.05 0.53	0.45	30	7.6	8	67	4.7	
7	30	0.2	0.002	—	2.0	0.43	0.05 3.13	3.23	207	92	12	331	7.8	
8	34	0.1	0.001	0	1.5	0.16	0.01 2.48	2.38	167	77	14	265	7.6	

and decreasing silica values with rising rates of rainfall on volcanic ash (Fig.30). TANADA (1951) pointed to increasing gibbsite content at the expense of kaolinite with increasing rainfall rate from 750 to 5,000 mm.

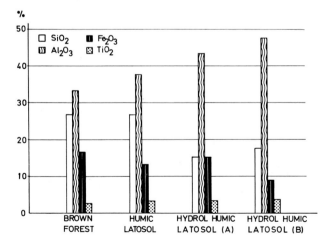

Fig.30. Relationship of rainfall rates, alumina and silica content in soils on volcanic ash from the Hawaiian Islands (after SHERMAN, 1958). Aluminium increases while silica decreases with increasing rate of rainfall.

Soft slopes in areas with high rates of rainfall, have a high rate of water infiltration, with subsurface conditions favouring the lateral movement of water. They are predestined to aluminium enrichment (Fig.31). Through decomposition

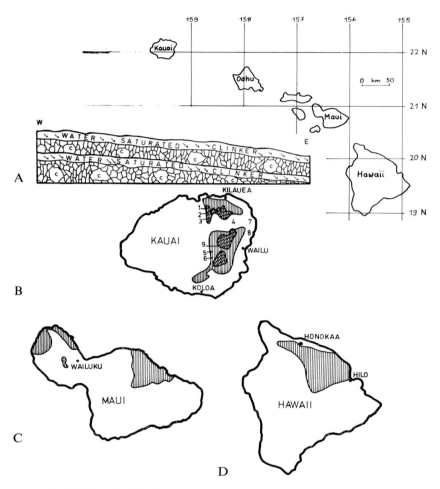

Fig.31. Hawaiian Islands

A. Basaltic flows with clinker at top of the lava, and cracks (c) supplying water to a lower klinker zone. The lavas dip or slope down from the volcanic ridge located along the centre of the island (W), where rainfall is greatest, to the coast (E).

B. Island of Kauai. Vertical lines: gibbsite aggregate (more than 10% gibbsite) in kaolin clay; diagonal lines: gibbsite in ferruginous bauxite; 1–8: wells for ground water analyses.

C. Island of Maui. Vertical lines: gibbsite in various forms, usually in pockets at various depths.

D. Island of Hawaii. Vertical lines: gibbsite–allophane clays developed on volcanic ash. (After ALLEN and SHERMAN, 1965; and SHERMAN, 1958.)

of the staghorn fern growing in great profusion on the northeast slopes, the soil is supplied with organic acids, thus increasing the speed of disintegration.

Because lateritization results in the dissolution and transport of nearly all the elements in the ground water, the catena pattern of either Fe-, Al- or silica-enrichment is governed by the direction and speed of ground water movement, hence by morphology.

Catena pattern. Many authors, in particular G. D. Sherman and co-workers (1949–1965) and ABBOT (1958), engaged in research on lateral differentiation of the latosol, and aluminium enrichments therein. SHERMAN (1958) described a catena pattern downhill from the Hawaiian Islands as: Halii series (upper slope) – Haiku series – Puhi series (bottom slope) (Fig.32).

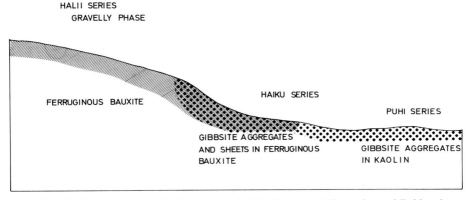

Fig.32. Catena rich in gibbsite composed of Halii series, Haiku series and Puhi series on the Hawaiian Islands. (After SHERMAN, 1958.)

Chemical composition. Chemical analyses of a weathering profile shows rapidly decreasing alkali, alkaline earths and silica values at a sharp boundary of basalt (A) and saprolite (11–13 m). There is progressive enrichment with iron, aluminium and titanium from the bottom towards the top of the profile. Chromium and vanadium also increase in the upper part (Table XVI).

SHERMAN (1958) also found lateral differentiation of soils on the islands of Maui and Hawaii. Extremely high rainfall on pitching lava flows, which provide excellent ground water circulation, led to enrichment of aluminium as lenticular bodies. Single layers or concretions consist of nearly pure gibbsite resulting from aluminium feed downhill.

The paragenesis of the latosols on the Hawaiian Islands changes both horizontally and vertically, depending on specific localities within the catena. Both chemical composition and paragenesis are governed by the direction of ground water movements and the degree of dehydration of the soils during dry seasons.

TABLE XVI

CHEMICAL AND SPECTROGRAPHIC RESULTS, IN PERCENT, OF ANALYSES OF SAMPLES OF WEATHERED BASALT FROM AN AUGER HOLE AND OF A SAMPLE OF ESSENTIALLY FRESH BASALT IN THE EASTERN PART OF KAUAI, HAWAII (after PATTERSON and ROBERTSON, 1961)

Depth (ft.)	Chemical analysis														Spectrographic analysis		
	SiO_2	Al_2O_3	Fe_2O_3	FeO	CaO	MgO	Na_2O	K_2O	H_2O	TiO_2	P_2O_5	MnO	CO_2	Sum	Cr	V	Ca
0–4	5.7	25.6	40.2	0.40	<0.10	0.65	0.04	0.12	18.6	5.6	0.27	0.12	<0.05	97[1]	0.16	0.069	0.08
4–10	8.6	25.5	39.1	0.37	<0.10	0.72	0.05	0.09	17.8	5.6	0.30	0.14	<0.05	98[1]	0.15	0.059	0.08
10–14	11.4	24.6	37.8	0.82	<0.10	0.91	0.06	0.05	17.2	5.7	0.36	0.20	<0.05	99	0.17	0.077	0.09
14–19	17.1	26.2	32.7	0.58	<0.10	0.83	0.06	0.03	15.8	5.1	0.31	0.25	<0.05	99	0.13	0.065	0.08
19–24	19.7	25.8	32.1	0.50	<0.10	1.0	0.06	0.03	15.0	4.9	0.32	0.27	<0.05	100	0.11	0.052	0.07
24–34	21.5	25.1	30.7	0.53	<0.10	1.0	0.06	0.03	14.6	4.9	0.36	0.27	<0.05	99	0.12	0.058	0.08
34–39	25.3	24.1	29.1	0.44	<0.10	0.87	0.06	0.03	13.9	4.5	0.38	0.27	<0.05	99	0.14	0.060	0.08
A[1]	45.0	11.8	4.3	9.3	9.7	12.0	1.8	0.62	3.1	1.9	0.30	0.26	0.09	100	0.064	0.031	>1.0

[1] Essentially fresh basalt from outcrop.

TABLE XVII

CHEMICAL COMPOSITION (IN %) OF FERRUGINOUS BAUXITES WEATHERED DIRECTLY FROM MELILITE–
NEPHELINE BASALT FROM THE WAILULA GAME REFUGE, KAUAI (after SHERMAN, 1958)

Sample	SiO_2	Al_2O_3	Fe_2O_3	TiO_2	H_2O^+
1	4.0	48.5	26.0	3.1	17.3
2	4.7	47.0	25.1	3.3	17.5
3	5.5	46.3	28.7	3.6	17.9
4	4.4	45.8	28.0	4.1	17.6
5	2.1	41.2	36.1	5.0	16.8
6	1.9	43.1	36.0	5.8	16.7
7	2.0	39.3	37.0	5.5	16.3
8	2.4	39.3	36.5	6.5	16.7
9	2.1	19.2	60.8	5.5	14.1
10	0.0	36.2	38.6	6.0	16.8
11	0.0	44.3	28.7	4.5	19.7
12	0.0	43.7	31.0	5.0	22.2
13	0.1	41.7	32.5	4.1	18.9

The ferruginous bauxites of the Halii series (upper slope) are characterized by very low SiO_2 and high Al_2O_3 and Fe_2O_3 contents (Table XVII):

The relic texture of the basalts withstood leaching to a large extent. There is successive decomposition of minerals resulting from differences in mobility of the elements participating in weathering. Alkalis, the alkaline earths and silica dissolve and are removed. The solubility of aluminium is lowest, causing aluminium hydroxides to reprecipitate quickly. "The hydrated aluminium oxides are stabilized in concretions or water stable aggregates. They lead to the development of horizons which are porous. The porous condition of the horizons develops conditions that would favour the formation of bauxite." (SHERMAN, 1958).

In the Haiku series there have been both the segregation of gibbsite and ferruginous aggregates and the precipitation of gibbsite into layers (sheets) (Tables XVIII, XX).

"In the soils of the Puhi series (bottom slope), the slower movement of the circulating water has slowed the mobility of the soluble silica to a point where resilification of gibbsite occurs in the clay-sized particles. Thus the kaolinite content increases as the circulating water movement slows down and as water available for the weathering processes decreases with a lower rainfall. In the Puhi soils, which are developed in areas of the lowest rainfall, there is evidence that iron oxide is accumulating more rapidly than gibbsite. The iron content of these aggregates is much higher than that found for aggregates on the steeper slopes. They have a low silica content. The coating of iron oxide of the surface will protect the gibbsite from resilicification" (SHERMAN, 1958; Table XIX).

TABLE XVIII

CHEMICAL COMPOSITION (IN %) OF BAUXITIC AGGREGATES AND GIBBSITIC SHEETS OF KAUAI SOILS
(after SHERMAN, 1958)

Bauxitic materials	SiO_2	Al_2O_3	Fe_2O_3	TiO_2
Ferruginous bauxite aggregates from Puhi soil	6.5	43.3	28.7	4.2
Ferruginous bauxite aggregates from Haiku soil	4.1	46.0	28.0	3.0
Gibbsite sheet white layer	1.2	58.9	1.2	1.8
Gibbsite sheet gray brown layer	12.8	52.0	12.2	3.6

TABLE XIX

CHEMICAL COMPOSITION OF A PUHI SOIL PROFILE ON MAUI ROAD JUST EAST OF THE ROAD ENTERING
KILAUEA VILLAGE, KAUAI (after SHERMAN, 1958)

Depth (inches)	Oxide analysis (%)					
	SiO_2	Al_2O_3	Fe_2O_3	TiO_2	MnO	H_2O
0–5	0.7	18.6	43.7	7.6	0.2	19.9
5–10	0.5	18.6	41.1	6.8	0.1	20.8
10–12	3.6	19.7	45.0	8.5	0.2	21.0
12–20	0.5	21.3	43.1	7.9	0.1	23.2
20–35	5.7	25.8	45.5	8.1	0.3	16.0
35–40	7.0	21.4	44.1	8.3	0.3	17.6
40+	7.6	24.7	42.0	8.3	0.2	16.2

TABLE XX

CHEMICAL COMPOSITION OF AGGREGATES FOUND IN THE GRAVELED HAIKU SOILS OF THE KILAUEA
PLANTATION, KAUAI (after SHERMAN, 1958)

Material	Portion of aggregate	Oxide analysis (%)			
		SiO_2	Al_2O_3	Fe_2O_3	TiO_2
Ferruginous aggregate #1	outer coating	1.7	16.0	56.5	4.7
	inner portion	0.4	27.0	43.8	6.9
Ferruginous aggregate #2	outer coating	0.4	18.6	57.0	4.6
	inner portion	1.1	26.2	44.1	5.0
Gibbsitic aggregate	outer coating	2.0	35.6	40.5	4.2
	inner portion	1.1	51.0	16.2	4.8

From the channel of the Wailu River on Kauai Island similar profiles with much aluminium enrichment on the upper slope by lateral feed with solutions are recorded by ABBOT (1958). Nearly pure bauxites formed lenses up to 15 cm in length or precipitated in cavities and fissures.

From areas with very high rainfall rates in the lower plains of the eastern part of Kauai, PATTERSON and ROBERTSON (1961) described weathering profiles 18–30 m thick with iron and silica enrichment maintaining relic textures: "The greatest resilicification of gibbsite to kaolin minerals occurs in soils of the very high rainfall area, where the internal drainage has been impeded so as to produce water-logged conditions of the weathering system." Finally there is development of a montmorillonite catena section on low ground which is extensively waterlogged.

Mineralogy. Mineral decomposition and neomineralization take place via solution and gel stages, partly in situ maintaining the basalt relic texture and partly through removal of solutions and reprecipitation in cavities, joints or as concretions.

The amorphous fraction. The amorphous fraction predominates in early stages of soil development. SHERMAN et al. (1964): "The amorphous fraction is extremely sensitive to changes in chemical and physical environments. Soils rich in amorphous hydrated colloidal oxides have a high water content, a low bulk density, and high cation exchange capacity. On dehydration these soils show an increase in bulk density and particle density and a loss of cation exchange capacity. The gels are rich in iron, aluminium and may also contain silica and titanium. Amorphous systems rich in aluminium and iron will, on dehydration, produce a separation of nearly pure gibbsite crystalline aggregates. Even in Al–Fe-gels pure gibbsite will separate with iron, forming its own system. The dehydration of iron-rich gels can lead to cryptocrystalline lepidocrocite → maghemite or goethite → hematite iron oxide systems."

Clay minerals. SHERMAN and UEHARA (1956) and SHERMAN et al. (1962) investigated the influence of a weathering environment on secondary mineral formation created by the supply and release of bases through leaching of olivine basalt.

In profiles with good drainage conditions, plagioclases weather to gibbsite, halloysite or kaolinite, depending on the concentration gradients of alumina and silica within the micro-environment (NAKAMURA and SHERMAN, 1965). Under heavy drainage conditions halloysite amygdules undergo desilicification along the periphery where acidic, silica-deficient water passes, attacking the halloysite by dissolving the silica. Halloysite is stable only if it is protected from such solutions, or if the solution passing by is saturated with silica. While alteration of feldspar to halloysite involves an increase in volume, a loss in volume follows desilicification of halloysite. This loss in volume is exemplified by the surface cracks clearly visible in the desilicificated halloysite. The decrease of halloysite is followed by an increase in gibbsite (UEHARA et al., 1966).

The transformation of olivine basalt, rich in iron, may result in nontronite formation in pockets and cavities. Montmorillonite clay was found in the lower layers of some profiles where bases and silica accumulated. In those profiles, kaolinite was found in the upper part and in the top layers.

Aluminium minerals. Gibbsite is the only aluminium mineral recorded. It formed in situ mostly in profiles of a higher terrain, partly from feldspar directly and partly via halloysite, indicating two stages of desilicification.

Aluminium is very mobile in solution; it may migrate downhill and precipitate as pure gibbsite in cavities or as concretions. Small gibbsite pisolites or bigger concretions grow from rhythmic precipitation, resulting in a shelly texture. ABBOT (1958) described nodular masses of gibbsite reaching 15 cm in length.

Boehmite has not yet been recorded from Al-laterites of the Hawaiian Islands.

Titanium minerals. Many papers deal with the weathering of titanomagnetite. According to MATSUSAKA et al. (1964), primary titanomagnetite of the basaltic and andesitic rocks seems to contain between 21 and 25 mole % TiO_2. Oxidation causes maghemite and anatase to form from titanomagnetite.

In Hawaiian soils, SHERMAN (1952) found concentrations of titanium dioxide ranging from 2.5 to 25.0%. The titanium dioxide content is higher in the surface horizon of soils formed in a climate of successive wet and dry seasons. The highest accumulation is recorded from humic ferruginous latosols, which occur in areas of successive wet and dry seasons but are adjacent to tropical forests. The conditions of accumulation of iron oxides and titanium dioxide are the same.

At or near the surface titanium dioxide is easily dehydrated to form concretions, coatings on aggregates or soil particles, or a massive horizon. In soils of the Haiku series titanium dioxide concretions may reach 0.5–1.2 cm in diameter. Titanium dioxide in the form of anatase rehydrates slower than iron oxide and is more resistant to reduction when the soils' internal drainage becomes poorer. Extremely low Eh-values reduce titanium dioxide, and it is leached from the soil. Certain conditions of poor drainage favour reduction and removal of oxides of iron and titanium by leaching, while gibbsite is resilicified to kaolinite.

Iron minerals. The proportion of amorphous iron hydroxides is sometimes very high. Depending on local pH and Eh conditions, iron ions or colloids may be transported over long distances. Fe^{2+} occurs in most cases as traces only in soils and is accommodated to a large extent by relic minerals such as titanomagnetite and magnetite. Lepidocrocite was detected only in one sample by X-ray techniques, but the mineral was identified repeatedly by DTA. Goethite and hematite are the most common iron minerals and are easy to identify by X-ray methods. They are formed either from local precipitation in situ or secondary enrichment in pisolites, concretions or crusts.

The pattern of mineral formation in Hawaiian soils is set out by Sherman as shown in Fig. 33.

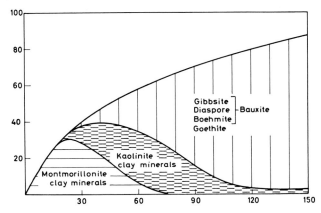

Fig.33. Distribution of clay minerals and bauxites in Hawaiian soil samples. Abscissa: annual rainfall in inches. Ordinate: percentage of clay and free hydrated oxides in the soil. (After SHERMAN, 1952.)

Ivory Coast

Equatorial Africa provides further examples of Late Tertiary ferrallitization. Investigations of DELVIGNE (1965) and SEGALEN (1965) on the Ivory Coast revealed the relationship of climate, morphology and drainage to specific soil formation. The fossil soils are considered to be of Late Tertiary to Pleistocene age. As both the continental coasts and the morphology have changed little since, climatic and soil zones are much the same nowadays. The morphology is governed by an old peneplain inclined to the south (north 350–400 m, south 50 m above sea level). There is a little more relief on the southern part, the mountains rising to 200 m above sea level. In the west only a few summits rise to 1,000 m. They are inselbergs made of granitic and noritic rocks, partly capped by old lateritic crusts. On the slopes there is latosol formation, which will be dealt with in this context.

There are two areas to be distinguished in view of the climate (Fig.34): the southwestern part from 5° to 8° latitude, with rainfall exceeding evaporation, and the northeastern part from 8° to 11° latitude, with evaporation exceeding rainfall (Table XXI).

Analogous to climatic regions one may define three zones of vegetation: the littoral zone, the zone of forests in the west and south, and the zone of light bush and savannah in the northeast.

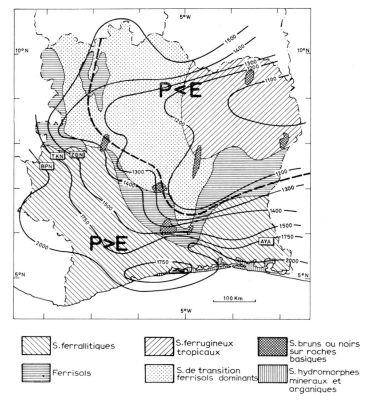

S.ferrallitiques

Ferrisols

S.ferrugineux
tropicaux

S. de transition
ferrisols dominants

S.bruns ou noirs
sur roches
basiques

S. hydromorphes
mineraux et
organiques

Fig.34. Combined map of climate and soils of the Ivory Coast. Numbers refer to rainfall isohyps with $P > E$ in the southwest, and $P < E$ in the northeast (P = precipitation, E = evaporation). Squares with numbers give localities of examined profiles. (After DELVIGNE, 1965.)

TABLE XXI

CLIMATIC ZONES IN EQUATORIAL AFRICA (IVORY COAST)

Climatic criteria	Climat guinéen forrestier (5°–8° latitude)	Climat sudanoguinéen (8°–11° latitude)
Mean annual temperature (°C)	24.5–27.2	24.5–28.2
Annual rainfall (mm)	1200–2000	NW 1700, NE 1100
Annual saturation deficiency (mm)	5–7	7–12
Number of months with rainfall	7–9	7–8
Number of wet seasons per annum	2 (May–July; Oct.–Nov.)	1 (June–Oct., maximum in August)
Number of dry seasons per annum	2 (extensive: Dec., April)	1 (Nov.–Feb./March)

There is a soil classification (Fig.34) corresponding to morphology and climate which distinguishes:

"sols ferrallitiques"

"ferrisols"

"sols ferrugineux tropicaux"

"sols de transition"

"sols bruns et noirs"

"sols hydromorphes"

From the ferrallitic zone of the southern and western areas with high rates of rainfall (Fig.34), DELVIGNE (1965) describes:

(1) A series of weathering profiles (TKN, ZGN, BPN), 1–3 m thick, on the slopes of Tonkoui/Ivory Coast. The environment is recorded as mountainous, the summit rising to 1,000 m above sea level. The source rocks are charnockite and norite, providing excellent drainage downhill. The vegetation is typified as "forrestier de montagne".

(2) Profiles (AYA) from the valley Bia north of the village of Ayame, Ivory Coast. They are several meters thick with well-marked vertical zoning. They developed from coarse-grained granodiorites and amphibolites of low quartz content. The climate and vegetation are similar to that of the Tonkui Mountain. Compared with previous profiles, these represent lower slopes where drainage intensity gradually becomes less during the weathering process. The lower slopes are covered by thick sheets of scree (Fig.35).

The combined profile series provides reconstruction of a simplified scheme of the catena on slopes with good drainage. All profiles are characterized by well-

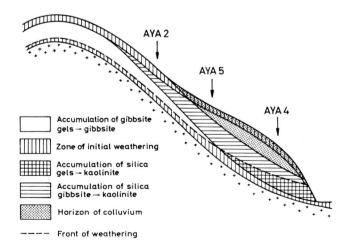

Fig.35. Catena in ferrallitic soil areas in the south of Ivory Coast. (After DELVIGNE, 1965.)

preserved relic textures, which rendered calculations by the method of MILLOT and BONIFAS (1955) feasible.

There is strong leaching of primary minerals in the weathering profiles, the solution being removed predominantly downhill. Secondary paragenesis occurs governed by both the quantity and chemical composition of the percolating ground water. The catena I, IIa, IIb and III is described by DELVIGNE (1965) as a function of morphology (Fig.36). In the top zone, zone I, strongly weathered rocks abruptly change into fresh rock within a few mm in vertical direction; at least 90% of the silica, the total amounts of alkalis, earth alkalis and aluminium were removed from the entire rock which consisted of Fe-Mg minerals.

Silica migrates from the zone in the monomeric form. Iron and alumina become relatively enriched here. The iron precipitates as amorphous hydrate which is later transformed into hematite or goethite, while aluminium crystallizes directly to gibbsite. In the middle and higher parts of the profile the secondary minerals consist of gibbsite and goethite only. The relic textures are very well preserved.

There is a slightly thicker weathering horizon at the boundary of source rock and soil profile in the middle zone, zone II, of the catena. This level, too, is characterized by removal of silica and bases, however silica migrates at a much slower rate. During feldspar decomposition, Al-Si gels are formed intermittently from which gibbsite crystallizes rapidly if all silica is removed. In the middle and upper parts of the profile of zone II, monomeric silica is removed by ground water. Portions of early gibbsite and silica react to form kaolinite. The primary texture is obliterated in most cases at this silicification level.

In zone III of the catena developed on the bottom slope, weathering and silicification overlap. Gels are widespread and are transformed into kaolinite via transient gibbsite. The silica solutions migrate at much slower rates, and only approximately 50% of the original silica content of fresh rocks is removed. All silica seems to be removed from primary lattices as polymers, while a certain amount of monomeric silica enters this specific level and replaces part of the silica dissolved in situ. Bases dissolve completely, but at much slower rates than at the top of the catena. The primary texture is preserved relatively well.

There is a vertical transport of solutions in addition to the ground water movement downhill. Part of the iron and aluminium content is dissolved from sections of the profile near the surface and removed by ground water. Aluminium rapidly crystallizes to secondary gibbsite or to kaolinite if monomeric silica is available. There is absolute enrichment with iron in the lowermost slope profiles where iron hydroxides or oxides may form a strongly impregnated zone.

Summary. In the uppermost zone, zone I, characterized by relative Al and Fe enrichment, gibbsite and goethite are the only secondary paragenesis. In zones II and III, characterized by increasing Al content combined with successive relative

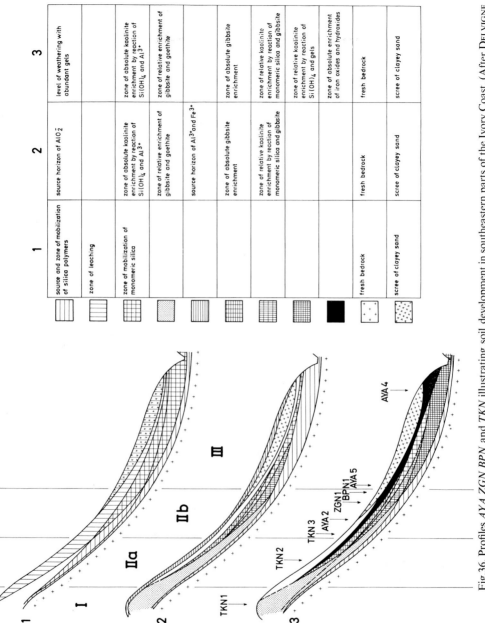

Fig.36. Profiles *AYA ZGN BPN* and *TKN* illustrating soil development in southeastern parts of the Ivory Coast. (After DELVIGNE, 1965.)

and absolute enrichment of silica, the mineral sequence is gibbsite–kaolinite–vermiculite.

There is also absolute enrichment of iron-forming hematite or goethite in parts II and III in the lower terrain of the slope. No desilicification of kaolinite to gibbsite was observed in either profile. The substantial proportion of gibbsite which occurs in specific sections near the surface of the lower slope profiles accounts for disintegration of Al minerals in the capping scree. The predominant pattern of transformation from primary into secondary minerals, was set out by Delvigne as follows:

Zone I + IIa	*Zone IIb*	*Zone III*
plagioclase–gibbsite	plagioclase–gel–gibbsite	plagioclase–gel–gibbsite–kaolinite
pyroxene–goethite	pyroxene–goethite	pyroxene–bowlingite–gel–goethite
		biotite–vermiculite–kaolinite

Southern India

Mineable bauxite formed on charnockites under certain conditions. Lenticular bauxite deposits several hundred m in length and 1–8 m thick developed on the slopes of the Early Tertiary peneplains in the Shevaroy Hills and in the Blue Mountains of southern India (KRISHNAN, 1942; VALETON, 1968). There is a transition from ferrallites with little kaolinite on the upper slope to siallites rich in silica on the lower slope. Relic textures show up folds of differently composed source rocks in the ferrallite zone.

Differences in the paragenesis of the folded series cause extremely irregular boundaries between fresh rock and weathered zone. Depending on the mineralogy, the change from fresh rock to bauxite or kaolinite clays is usually rather abrupt. There is no distinct vertical zoning of the bauxite profiles with an upper concretionary zone and iron crust developed at the top.

Mineralogy and chemistry. The common neomineralization is kaolinite in the fine-grained micaceous source rock and gibbsite in the coarse-grained primary rocks rich in feldspars. During the initial weathering phase a thin film of gibbsite or goethite developed along fissures on the minerals weathering readily, such as feldspars, augites and garnet (Plate IV,17). Secondary hematite and goethite precipitated from iron supplied by augite and garnet. They reflect detailed contours and fissure systems of augites and garnets, later dissolved leaving behind a cavity pattern (Plate V, 20, 21). Gibbsite may develop a similar fissure skeleton replacing feldspar (Plate V, 19). In other places during feldspar transformation Al precipitated in situ to form gibbsite which burst through the original grain contours due to an increase in volume and swelling.

Relic quartz (maximum up to 50–90%) is fringed by corrosion and dissolution to form angular grains (Plate V, 18). Apparently there was a high mobility of aluminous solutions and strict control of silica removal causing quartz to be dissolved to a large extent and providing space for gibbsite precipitation under specific local conditions of the Early Tertiary weathering cycles.

The porosity of the aluminous rocks ranges from 30 to 50%. Only in certain part of the pore space coarse-grained gibbsite crystallized during the periodical supply of Al solutions. The chemical composition of the bauxites is given in Fig.37.

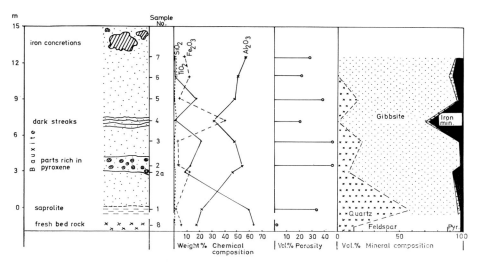

Fig.37. Chemical and mineralogical composition of bauxite profile Y-II on charnockite; Shevaroy Hills, southern India (VALETON, 1968).

The triangular projection of SiO_2–Fe_2O_3–Al_2O_3 (Fig.38) demonstrates the development of a saprolite and allite zone initially along the line of equal Al/Fe ratios. For the most part there is dissolution and removal of silica only, causing relative enrichment of aluminium and iron in skeletons. In irregular iron-rich parts there is additional impregnation with iron that precipitates from permeating solutions.

The bauxite developed on charnockites in southern India corresponds in principle to the ferrallite type of Hawaii and the Ivory Coast:

the dominant textures are relic textures of rocks weathered in situ;

iron and aluminium become enriched relatively; the lateral supply of solutions that resulted in Fe and Al impregnation is subordinate;

in contrast to plateau bauxites, this type does not develop enriched zones of aluminium hydroxides with pisolitic textures or capping iron crusts;

these bauxites do not form boehmite or diaspore; besides hematite goethite is a common primary mineral.

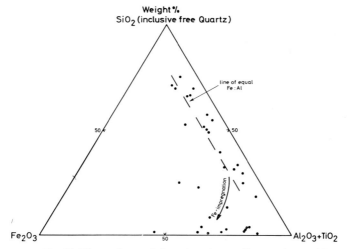

Fig.38. The analyses of three bauxite profiles on charnockites plotted on the graph show that there are constant Al/Fe ratios of bauxite and source rock or that iron impregnation took place.

Plateau type

The most important deposits of this type occur on rocks of the Precambrian basement, on syenite massives, andesites and on plateau basalts, as shown on the attached map. These deposits differ widely in size and economic importance. There is no uniform nomenclature. Ground water laterites (American references in general), plateau laterites (Russian literature in most cases), and high-level laterites (mainly Indian references) are the terms generally applied to these types. PATTERSON (1967) refers to a "blanket type". These are coverings of laterite 10–40 m thick, depending on the source rocks, but which occur regionally on specific plateau levels only in present tropical and subtropical areas (Fig.39).

They developed mainly during Early Tertiary time and became fossilized since. There is a distinct vertical zoning, reflecting specific mineralogical composition, sometimes with considerable Al enrichment as flat lens-shaped bodies, forming large bauxite deposits in several cases (Brazil, Guiana countries, Equatorial Africa, Southeast Asia, Australia).

Plateau type on basic rocks

The source rocks of the bauxites on plateaus of the Deccan peninsula are basalts, mostly of tholeyitic composition, which formed from viscous lava flowing over large areas with flat escarpment. The even plateaus are very extensive. There are no elevations above this plateau. Fox (1923, 1927) was the first to note facies differentiation of plateau laterites in India. He theorized a predominantly lateral

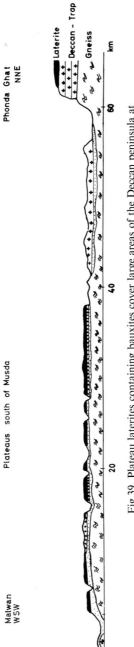

Fig.39. Plateau laterites containing bauxites cover large areas of the Deccan peninsula at specific levels. They are fossil, and were destroyed by young erosion following structural displacement. (After FOOTE, 1867.)

transport of Al, Fe and Si solutions as a function of morphology. Unfortunately development of the high-level plateau laterites was dated as Recent by him, which later caused much confusion.

A decisive step forward was made by regional geological mapping of the high-level laterite facies by Roy CHOWDHURY (1958).

The formation of the fersiallite-ferrallite-allite catena was governed solely by the flatness of the country in Early Tertiary times and the corresponding drainage pattern. The distribution of fersiallites and allites on the 1,000-m level plateaus of the Deccan peninsula bears no relationship whatsoever to Recent edges of inselbergs.

In high-level areas there was a rather distinct relationship between well-drained sources and upper courses of Early Tertiary rivers flowing northwest and allites, and between water-logged sources and upper courses of similar rivers and

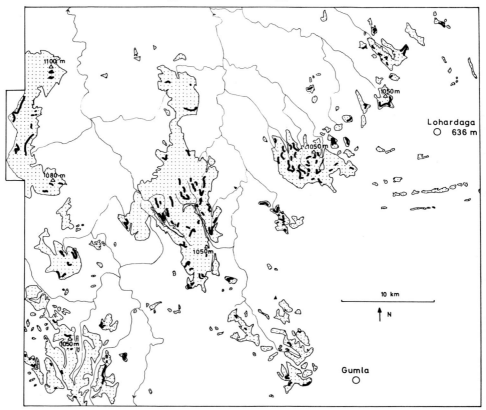

Fig.40. Laterite distribution on the high-level plateaus of the Bihar Mountains, India. There is a facies change from laterites (dots) formed in situ on basalts into bauxite (black lines), related to Early Tertiary drainage systems directed northwesterly. The topography of the plateaus was developed in a morphological low position. (After Roy CHOWDHURY, 1958.)

siallites (Fig.40). Explorations made by the aluminium industry proved regional extension and thickness of bauxite lenses within the laterite cover to be a function of the density of fossil rivers and their rate of flow. It follows that laterization was in progress on an undisturbed peneplain which was extensive but at low levels. All laterites of the whole Deccan Plateau are considered to be of approximately the same age. They predate Neogene tectonic movements which caused their displacement. They are older, too, than many valley systems and cuesta scarps covered by scree (VALETON, 1967a; Fig.41). The same facies pattern is always observable in many laterites with high alumina content in India.

Primary facies with subordinate postdiagenetic changes: Gujerat, India

Allitic and fersiallitic facies developed from a laterite belt in southwestern Kathiawar on a peneplain at sea level in the north and south, respectively. The valley systems cut into the peneplain are confined to a maximum depth of 20 m. Both relief and paragenesis of these laterites and bauxites became fossil under cover of postbauxitic marine Tertiary sediments (Nummulithicum/Eocene). The laterites and bauxites are probably of Early Eocene age.

Two profiles, approx. 500 m apart, are distinct members (Fig.42) of a continuous series from fersiallites of formerly central areas of the plateau (M II) and allites on a fossil valley slope (M I). There was transformation in situ maintaining both macro- and microtextures, which means that the volume remained constant (Plate I, 1; III, 12). The vertical and horizontal facies differentiations are given in Fig.43 and Table XXII.

The respective lower fersiallite and siallite saprolites are distinguished by highest porosity and excellent preservation of basalt relic textures. Relative enrichment processes resulted in secondary mineral formation as follows:

	fersiallite saprolite under laterites	*siallite saprolite under bauxites*
primary minerals	inner part of the plateau	valley slope
olivine	serpentine–goethite	kaolinite–goethite
augite	goethite	(goethite)–kaolinite
feldspar	kaolinite or gibbsite	(kaolinite) or gibbsite
glass	kaolinite	kaolinite

The fersiallitic saprolite is richer in goethite and kaolinite, and there are traces of serpentine in contrast to the siallitic saprolite, with little goethite but a great deal of gibbsite in addition to kaolinite.

An abundance of aluminium and titanium in the allite zone of both profiles destroyed relic textures, and spongy, porous gel textures of gibbsite composition developed (Plate II, 8). Because smaller amounts of solutions were supplied to central parts of the plateau and because of more uniform water impregnation, the regular spongy textures fossilized. But along valley slopes where humidity fluctuated seasonally widely and the rate of penetration of groundwater solution was high,

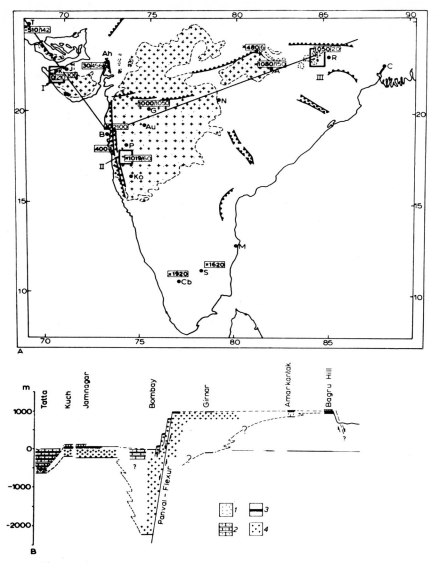

Fig.41. The Deccan peninsula.

A. Structure and regional distribution of Trap basalts (after HAZRA and RAY, 1962).
Bauxites investigated: *I*. Gujerat (ref. to text); *II*. Udagiri Plateau; *III*. Bihar Mountains; *Cb* = Coimbatore, bauxite on charnockite of Kotagiri; *S* = Salem, bauxite of Yercaud. The boldface numbers (m) give the altitude of the basalt surface and associated high-level laterites, while light printing refers to basalt thickness; ++ = Trap basalts.

B. Schematic section through the peninsula from Tatta to Ranchi (strongly exaggerated), illustrating structural displacement of the basalt surface covered by high-level laterites. The laterites are overlain by marine strata (Nummulithicum and younger) in the west and by eolian sediments in the east.

1 = eolian sediments of Bagru Hill/Bihar Mountains; *2* = marine Tertiary strata; *3* = bauxite laterites; *4* = Deccan Trap basalt. (After HAZRA and RAY, 1962, supplemented by VALETON, 1967a.)

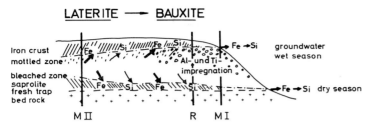

Fig.42. Relationship of vertical and lateral facies differentiation of laterites in the district of Halar, Gujerat, and drainage. Investigation of three profiles showed highly ferruginous laterites in formerly central parts of the plateau and Al-rich bauxites along valley edges.

rhythmic shrinkage and precipitation resulted in fissures, flow textures, and piso-lite formation (Plate III, 9, 10). The cores of the pisolites consist of spongy bauxite, while shells are made of boehmite. In transitional zones of spongy and pisolitic

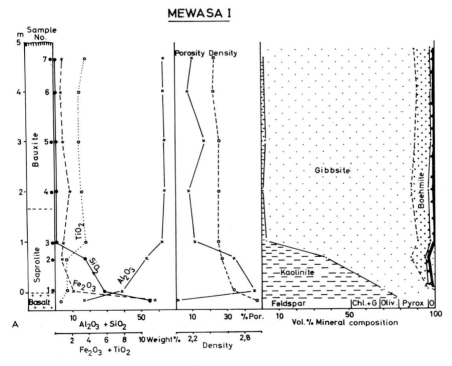

Fig.43A, B, C. Chemical composition, porosity and density and mineral composition in profiles from central parts of the plateau (Mewasa II); close proximity to valley edges (Mewasa I) and intermediate site. Note high porosities and high kaolinite content of the lithomarge (= saprolite). The carbonate content is a result of later marine transgression.

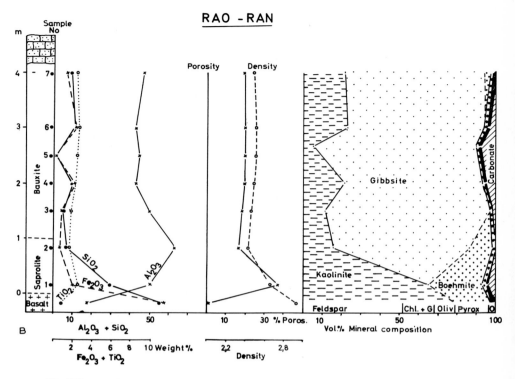

Fig.43B.

bauxite either texture developed depending on the textures of the former basalts which governed drainage and transformation. Accordingly, there were two distinct stages of mineral formation, i.e., gibbsite formation followed by boehmite. As the valley slope was approached, absolute enrichment of Al and Ti increased with intensified pisolite and boehmite formation with a marked decrease in porosity.

The upper lateritic crust is strongly developed in central parts of the plateau but is missing along valley slopes. The iron crust is fersiallitic in the upper but ferrallitic in the lower section. The texture is vesicular, with a hematite skeleton filled by kaolinite and gibbsite in the upper and lower parts, respectively (Plate I, 2; II, 6). There was secondary goethite formation at the expense of hematite (see below). The lower part of the iron crust is occasionally pisolitic.

Basaltic relic textures are still easily observable. However, they become obliterated where absolute iron enrichment increased. This zone is characterized by a sharp division of Fe on the one hand and Al or Al + Si on the other. Isomorphous replacement of Fe by Al in the minerals hematite and goethite was subordinate.

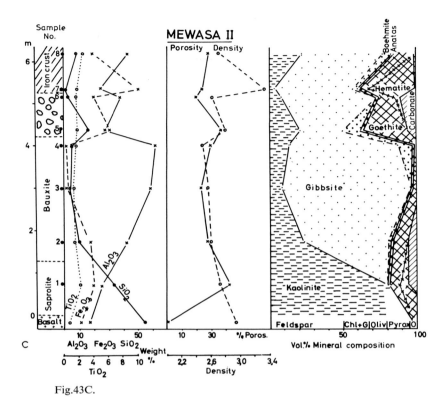

Fig.43C.

TABLE XXII

VERTICAL AND HORIZONTAL FACIES DIFFERENTIATION OF GUJERAT PROFILES

Morphology	Plateau (M II)	Valley slope (M I)	Valley floor
	fersiallites	allites	multicoloured kaolinite clay
Profile subdivision:			
lateritic iron crust	thick iron crust	thin or missing iron crust	
Mottled zone	mottled zone		
Pallid zone	bauxite = allite	bauxite = allite	
Saprolite (lithomarge)[1]	red fersiallitic saprolite	white siallitic saprolite	multicoloured clays
Source rock	basalt	basalt	basalt

[1] Saprolite is referred to as lithomarge in India.

The dominant aluminium minerals are gibbsite and boehmite with traces of diaspore. The *d*-values do not suggest isomorphous replacement of Al by Fe. However, aluminium and iron minerals are intimately intergrown both in the matrix and in the shells of pisolites in the upper transitional zone from allite to ferrallite. This transitional zone is only a few centimetres or decimetres thick in most cases. The aluminium–iron separation becomes even more distinct towards the top with increasing distance from the allite zone.

The laterization with constant volume allows the calculation of chemical balance. Silica removal resulted in not only relative but also absolute enrichment of Fe, Al and Ti with distinct differentiation in both the vertical and horizontal direction (Fig.44). Because there is 150–350% excessive Al and Ti compared to the source rock in both profiles M I and M II, a horizontal flow of solutions rather than a vertical flow was the cause of such concentrations. The distance over which solutions migrated was in the order of several kilometers at least.

The triangular projection of SiO_2–Fe_2O_3–Al_2O_3 demonstrates that the rocks of different zones correspond to specific fields and hence are defined by their chemical composition. It becomes obvious that SiO_2 was removed from all rocks paralleled by Fe dissolution in saprolite, while Al impregnation was confined to the allite horizon and iron impregnation to the ferrallite zone and the upper crust (Fig.45).

The irregular upper boundary of the profiles and the reworked material spread over the surface indicate denudation. The uppermost 20–50 cm of the profiles show a Recent greyish-brown weathering zone superimposing the lateritic profile. Therefore, the laterite profiles are rudimentary. The upper profiles underwent reworking partly during Tertiary transgression and partly in Recent time.

Secondary facies developed from postdiagenetic mineral transformations: Deccan high-level plateaus

In principle all laterites of the high-level plateaus are identical with the Al laterites in Gujerat, which in vertical and lateral facies change from fersiallites in central parts of the plateaus to allites along valley slopes. Examples from the western Ghats and the Bihar Mountains (ROY CHOWDHURY, 1958) prove that the facies differentiation is in no case related to Recent morphology; however, there is a connexion with Early Tertiary rivers on the plateaus.

The fossil character is stressed by:

reworked material and scree on plateau surface and escarpments, respectively;

Early Tertiary fans on foreland plains that contain reworked material of the high-level plateau laterites;

younger eolian sands on plateau surfaces and Recent soil formation.

Therefore, laterization predates valley deepening and was in progress at a lower level.

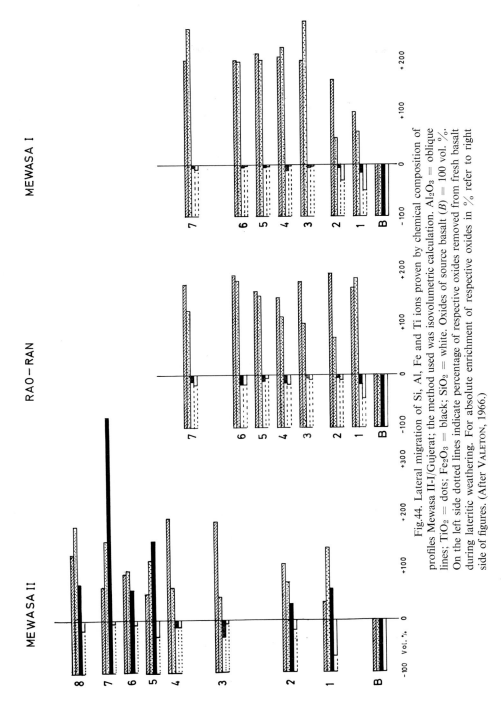

Fig.44. Lateral migration of Si, Al, Fe and Ti ions proven by chemical composition of profiles Mewasa II-I/Gujerat; the method used was isovolumetric calculation. Al_2O_3 = oblique lines; TiO_2 = dots; Fe_2O_3 = black; SiO_2 = white. Oxides of source basalt (B) = 100 vol. %. On the left side dotted lines indicate percentage of respective oxides removed from fresh basalt during lateritic weathering. For absolute enrichment of respective oxides in % refer to right side of figures. (After VALETON, 1966.)

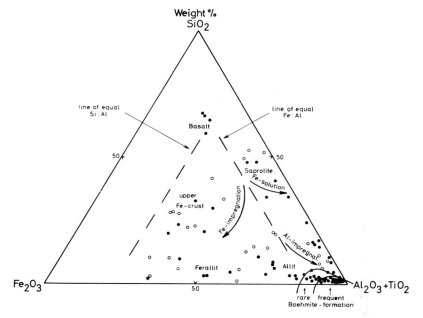

Fig.45. Triangular projection of chemical composition of individual zones of bauxite profiles: basalt–saprolite–allite–ferrallite–capping iron crust. There is clear separation of the various zones, and the removal of Al or Fe or impregnation with these elements may be read from the diagram; boehmite formation is confined to areas of strong aluminium impregnation.

Likewise corresponding Al laterites in Gujerat on high-level plateaus developed three facies governed by drainage conditions of the Early Tertiary relief (Fig.46):

in central areas of the plateau, laterites rich in Fe and Si;

on fossil river slopes bauxites rich in alumina;

on the edges of flat fossil sources with slow rates of drainage, kaolinite strata rich in Al and Si.

The polygenetic development resulted in considerable changes of mineral composition and texture. The transformations took place in an oxidizing environment with excellent drainage conditions and in part caused development or degradation of the laterites through surface weathering, depositional and leaching processes.

(*1*) Laterites rich in Fe and Si from central areas of the plateau. In the transitional zone of the basalt and weathering profile, e.g., in the *saprolite*, leaching was very intense, and two phenomena may be observed (Fig.47):

In places dissolution removed at least 50–66% of the saprolite and resulted in the complete breakdown of the textures (Plate III, 11). Kaolinite and iron

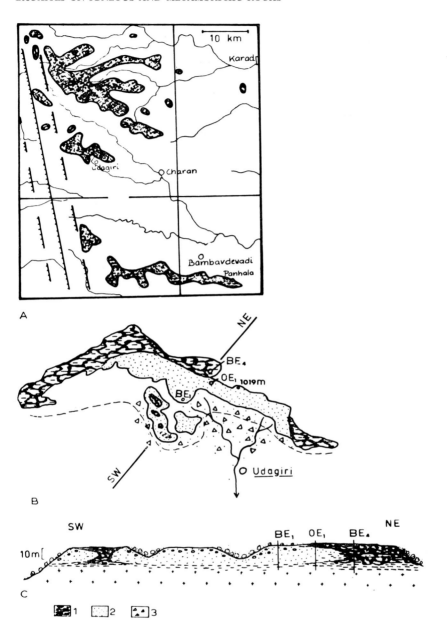

Fig.46. Udagiri Plateau. A. Geography of the plateau in the western Ghats (India) covered by aluminous high-level laterite. B., C. Map and vertical section showing distribution of laterite-bauxite facies pattern (horizontal lines = saprolite; crosses = basalt).
1 = iron-rich laterite; 2 = bauxite; 3 = bauxite scree. (After VALETON, 1967a.)

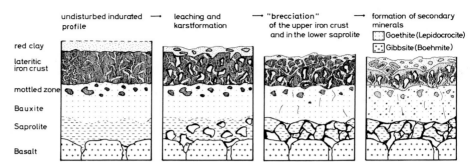

Fig.47. Mode of polygenetic alterations of high-level laterites on basalts in India. *1.* Lithification of laterites following rise above the water table; *2.* leaching, particularly intensive in the iron crust and in the lithomarge (saprolite), development of cavities: karst formation; *3.* formation of breccias through continuous leaching in the iron crust and in the lithomarge; *4.* secondary transformation of the hematitic crust into goethite (lepidocrocite), recrystallization and neomineralization of gibbsite (boehmite, diaspore) in the bauxite zone.
Processes *2–4* interact and provide sustained relative aluminium enrichment. (After VALETON, 1967b.)

minerals dissolved, while gibbsite with a spongy texture proved to be the most stable mineral, undergoing only a breaking apart into more or less angular fragments. These settled in a soft red matrix that consists of kaolinite, goethite and hematite, and developed a basal breccia. Further leaching of such facies exposed by younger erosion provided for cavity and tunnel systems with periodic subsurface water courses. The extent of leaching and washout varied with the amount of present-day exposure of the profiles.

In the middle *zone of allite* there was essentially recrystallization of gibbsite. The texture became coarse-grained with gibbsite crystals up to 100 μ in size (Plate IV, 15, 16). This gibbsite is triclinic in most cases. There is irregular extinction of the crystals caused by stunting traces of hematite. Sometimes a breakdown of the rock is observable (Plate II, 7), and a breccia-like texture results.

All profiles of the *overlying lateritic iron crusts* may be subdivided into lower and upper zones that are hard and soft, respectively. They are separated by a regular undulating boundary.

The iron-rich components have an irregular vesicular texture with wide meshes, a kind of network of iron concretions that touch in this laterite crust. The light minerals rest in between, as veins or cavity fillings. In the lower part of the iron crust gibbsite predominates. It recrystallized and formed a coarse-grained matrix. For these reasons the rock hardened extensively and became resistant to weathering.

The upper part of the lateritic iron crust is lighter in colour but made of the same vesicular texture. The light-coloured minerals belong mainly to the kaolininite group and form a soft matrix. Morphological exposure during wet seasons governed the formation of the matrix. On plateaus with poor drainage the layer silicates acted as water-logging horizons. Portions of the hematite disintegrated with decreasing crystallinity towards the surface, favouring reprecipitation of goethite and lepidocrocite. Along valley edges or cuestas with good drainage the kaolinite matrix was washed from the hematite-goethite skeleton to leave behind an extremely porous relic texture which collapsed in time. The loose debris were reworked as ironstone pebbles (Canga in many profiles of Africa; Plate I, 3–5). The secondary precipitation of Fe, Al and Si in pore space was minor in this case as cavity fillings or thin films of hematite, gibbsite and kaolinite are extremely rare. The areas of these high-level plateau laterites, rich in Fe and Si, are characterized by:

breccia texture caused by leaching at the base and top of the profiles;

pore space devoid of secondary precipitates;

recrystallization of the aluminous matrix to coarse-grained gibbsite;

a high percentage of kaolinite in the upper profile sections;

traces of boehmite, diaspore and lepidocrocite in the topmost and lowermost profile sections;

hematite replacement by secondary goethite;

relative enrichment of the allite component of fersiallites by disintegration of hematite and washout of kaolinite and goethite.

(2) Aluminous laterites on valley edges with good drainage. The soft *saprolite* zone (lithomarge), one to several metres thick in most cases, is the most unstable zone of these profiles; usually it is very porous and light yellow or grey coloured, indicating a low content of primary iron. The aluminous parts recrystallized to form a hard stable skeleton of coarse-grained gibbsite, from which the soft kaolinite was washed out. There was very little breccia formation because of primary higher aluminium content and hence greater stability.

A substantial part of the profiles along valley edges consists of a light greyish to pink-coloured allite zone several metres thick.

The dominant texture is porous and spongy with very coarse gibbsite crystals. Laterally, in the direction of better drainage, it passes into pisolitic bauxite. The cores of the pisolites are made of spongy gibbsite matrix, surrounded by boehmite shells. Depending on the density of Early Tertiary river systems and rate of flow, the areas of pisolitic boehmite differ widely in size, e.g., from 10–100 m with a maximum of several kilometres in diameter. This paragenesis is very stable and suffers only very small changes from later weathering on the high-level plateaus. These are:

increase in porosity;

partial filling of pore space by secondary gibbsite;

recrystallization of the matrix to form coarse crystalline gibbsite;

traces of diaspore formation besides gibbsite and boehmite;

partial disintegration of hematite and replacement by goethite.

(*3*) Siliceous facies of areas with poor drainage. The profiles have approximately the same or somewhat greater thickness than normal profiles. They formed in fossil depressions or sources with poor drainage. Apparently drainage was just sufficient to mobilize and remove iron. The very low iron content is confined to goethite. Thus a soft clayey siallite without free aluminium developed in both the lower and upper profile sections; hence it is nearly pure kaolinite clay. In the middle profile sections, gibbsite is the dominant mineral after kaolinite. The gibbsite forms either a hard skeleton or concretions in a soft kaolinite matrix. Such relatively soft siallite profiles rapidly wash out and erode if exposed.

The polygenetic changes of the aluminous laterites of the high-level plateaus yield the following results:

(*a*) The most stable members are allites with spongy or pisolitic textures, with low iron content. They form well marked outcrops of hard rock.

(*b*) Since pisolitic allites usually formed the edges of Early Tertiary valleys, retrogressive erosion may be calculated from debris of such facies on recent slopes or in screes.

(*c*) The siallites of areas with impeded flow are soft and brittle and are quickly destroyed with relative enrichment of the allite component.

(*d*) There is very strong leaching of Si and Fe in the fersiallites, particularly in the basal iron-rich kaolinite zones, which also means relative enrichment of the allite component.

(*e*) Diagenesis of aluminous laterites on basalt proved silica to be the phase of highest mobility. It is followed by iron, which is also removed from the edges quantitatively. Aluminium and titanium, however, concentrate along the edges of the plateaus to form bodies several 100 metres to several kilometres in length.

(*f*) Younger polygenetic transformations of the highly aluminous plateau laterites proved allites to be the most stable facies, which are preserved as monadnocks practically unaltered. The siallites and ferrallites are relatively enriched with the allite component by leaching of the soft kaolinite phase. Fine-grained detrital material of iron minerals and kaolinite was frequently transported over long distances and deposited in depressions.

Australia

Al laterites on basaltic rocks which have been extensively investigated are the bauxites occurring in Queensland, New South Wales, Victoria and Tasmania. They are fossil and partly covered by lignitic sediments and younger basalt flows (New South Wales), displaced by younger tectonic movements. They are an example of polygenetic transformation of bauxites under acid and reducing conditions.

At many places they have been eroded considerably and occur as truncated profiles over large distances. Most deposits can be dated Eocene or Oligocene (OWEN, 1954). The Al laterites are very similar to those on the Trap basalts in India.

The primary basalt textures are well preserved, too, in most cases, and indicate weathering in situ, maintaining constant volume.

The laterite blanket formed on a slightly undulating peneplain and normally 7 m thick, reaches 17 m in extreme cases.

The upper boundary is very regular, but there may be strong irregularities of the lower boundary because of different textures of source rocks (dense basalt, tuff, agglomerates).

Both lateral and vertical facies differentiation were a function of the same laws which controlled formation of the aluminous laterites of India and resulted in the same zoning of paragenesis and respective laterite textures.

In general, the bauxite lenses within the laterites are small (Fig.48A; Table XXIII). In Victoria and Tasmania denudation is often at such an advanced stage that only monadnocks of bauxite bodies resistant to weathering remain.

The aluminium-rich zones of the bauxites are also highly ferruginous (10–30% Fe_2O_3), and are therefore regarded as typical mottled zones. In the mottled zone there are gibbsite concretions in a soft clay matrix and earthy rocks in the lower and upper sections, respectively. Several profiles given in Fig.48b demonstrate vertical zoning.

Various deposits, such as that at Ouse in Tasmania, show that basalts do not always overlie laterites directly, but that there are intercalations of lignitic sands and clays. They indicate a rising water table and prove water-logged conditions. There was a mobilization of silica in overlying sediments which attacked the pisolite zone and caused the gibbsite matrix to react to kaolinite (Fig.49). In other parts "pale blue hard kaolinitic clay mainly developed along fractures as veins and irregular masses" (OWEN, 1954). The clay minerals are kaolinite and halloysite. In a reducing environment iron is attacked by humic acid, dissolved and reprecipitated as coarse crystalline siderite in veins and cavities. The siderite content may reach 18%.

In other bauxites with overlying lignites some pyrite was formed in addition to siderite crystallization (e.g., Buln Buln, Victoria). Such bauxites are greyish in colour, and the iron content often exceeds that of primary bauxites because of secondary mobilization and iron supply.

In the laterites of St. Leonards, Tasmania, high-quality bauxite developed from highly ferruginous laterites by secondary dissolution and removal of iron from the iron crust of the concretionary zone. Iron was either removed or reprecipitated as siderite in the lower saprolite. The relative enrichment of aluminium minerals in the upper and lower profile sections rendered the bauxite mineable.

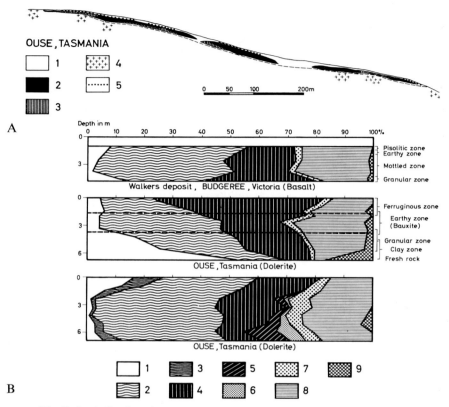

Fig.48. Australian bauxites.
A. Bauxite at Ouse, Tasmania; remnants of formerly extensive laterite covers on dolerite.
1. overburden; *2.* bauxites of economic grade or better; *3.* bauxites below economic grade; *4.* dolerite; *5.* bauxite boulders in clay.
B. Chemical composition of various bauxite profiles on basaltic rocks in Victoria and Tasmania. *1.* SiO_2; *2.* free Al_2O_3; *3.* fixed Al_2O_3; *4.* Fe_2O_3; *5.* FeO; *6.* CO_2; *7.* TiO_2; *8.* ignition loss; *9.* not determined. (After OWEN, 1954).

Plateau type on variable rock types

Arkansas, U.S.A.

Further examples of polygenetic transformations of primary bauxites under water-logged conditions are the bauxites capping the nepheline syenites in Arkansas, U.S.A.

In Arkansas, U.S.A., the Palaeozoic sedimentary basement is penetrated with several nepheline syenite domes. They are surrounded by clayey, sandy sediments of the Paleocene Midway Group. In Eocene times, development of the undulating Gulf Coastal Plain was in progress on this rock group. It includes huge bauxite deposits which are covered by the Eocene Wilcox Group.

TABLE XXIII

ZONES IN THE LATERITE PROFILES OF AUSTRALIA (after OWEN, 1954)

Zones			Common thick-ness (ft.)	Analysis No.	Typical composition				Deriva-tion[5]
LACROIX (1923)	WALTHER (1925)	Ideal bauxite section			SiO_2 (%)	Al_2O_3 (%)	Fe_2O_3 (%)	Free Al_2O_3 (%)	
Concretionary (zone de concretion)	ferruginous	derived soil	1–5	1	4.0	35.5	35.2	29.8	BT
				2	9.3	31.3	34.1	21.0	BI
		pisolitic[1]	4–20	3	3.3	52.4	13.8	49.7	SW
				4	2.5	42.7	31.6	33.7	BI
				5	3.6	38.5	36.4	27.9	BM
				6	2.9	28.0	50.0	—	BT
		tubular and massive[2]	4–8	7	2.5	48.4	19.6	47.8	SW
				8	4.6	34.2	36.9	32.4	BM
				9	1.3	26.7	47.1	25.8	BM
				10	3.4	38.7	33.2	36.5	DO
	mottled (Fleckenzone)	earthy: with or without poorly preserved relic texture, varie-gated	10–20	11	3.9	37.9	32.1	32.3	BM
				12	3.8	55.3	4.2	50.3	BM
				13	2.4	44.9	24.7	42.7	DO
Leached (zone de départ)		nodular: hard gibb-sitic nodules in soft clayey matrix[3]		14	8.0	38.7	27.0	32.7	BT
				15	9.1	36.9	28.5	29.6	BK
				16	11.1	48.3	13.5	37.8	DO
				17	9.7	40.4	23.8	30.1	DL
		granular: relic texture well preserved	5–10	18	18.3	25.5	39.9	7.2	BM
				19	3.7	43.8	26.5	40.6	DL
				20	12.8	39.2	22.4	28.1	DL
	pallid (Bleichzone)	clay zone	0.5–40	21	30.5	35.5	12.5	11.2	BM
				22	24.0	23.2	34.2	—	DL
				23	30.3	28.7	23.7	—	DL
		parent rock[4]	—	24	64.6	19.3	2.5	—	SW
				25	45.6	14.8	11.7	—	BI
				26	53.4	14.7	10.3	—	DL
				27	54.8	14.5	11.3	—	DO

[1] Analysis 4 represents 24 ft. of pisolitic laterite at Emmaville, N.S.W. The pisolitic zone in Wessel Islands ranges up to 16.5 ft. thick.

[2] Tubular laterite is not well developed in Tasmania. Sample No.10 is a massive ferruginous bauxite containing a few poorly developed pisolites with earthy cores.

[3] Nodular zones are uncommon and poorly developed in Australia.

[4] Total iron in these rocks is expressed as Fe_2O_3.

[5] Explanation of letter symbols: B = basalt; D = dolerite; S = micaceous siltstone; I = Inverell area, N.S.W.; K = Kingaroy, Qld; L = St. Leonards, Tasmania; M = Moss Vale, N.S.W.; O = Ouse, Tasmania; T = Midlands, Tasmania; W = Wessel Islands, N.T.

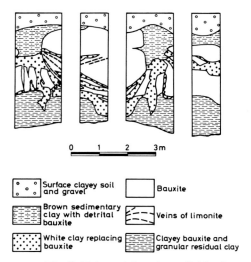

0   1   2   3m

° ° ° Surface clayey soil and gravel	Bauxite
Brown sedimentary clay with detrital bauxite	Veins of limonite
White clay replacing bauxite	Clayey bauxite and granular residual clay

Fig.49. Epigenetic bauxite resilicification resulting in kaolinite formation along fissures and joints. Limonite veins penetrate the bauxite, overlain by brown sedimentary clay with bauxite. There is evidence of erosion at the bauxite surface prior to clay sedimentation. (After OWEN, 1954.)

The bauxite deposits are subdivided into:

(*1*) Discontinuous blanket-like deposits over the weathered surface of the nepheline syenite hills and as lenticular deposits resting unconformably on the gently sloping surface of the Wills Point Formation near the break in the slope where it comes into contact with the igneous rock.

(*2*) Secondary deposits in stratigraphically higher units. The bauxite deposits on nepheline syenite formed and transformed in several phases: (*a*) subsurface weathering of nepheline syenite in an oxidizing environment with conservation of granitic relic textures of the leaching zone in the lower profile section and pisolitic textures in the upper concretionary zone; (*b*) secondary drainage and neomineralization resulting from deeper erosion by rivers and sinking of the ground water table; (*c*) epigenetic mineral transformations in a reducing environment under lignitic fresh-water sediments; (*d*) Recent weathering in an oxidizing environment.

The residual deposits on nepheline syenite nowadays are lens-shaped blanket deposits a few centimetres to 20 m thick, the length of which ranges from several metres to several kilometres. GORDON et al. (1958) describe the relationship of relief and bauxite formations as follows:

"Bauxites formed along ridges and slopes in the minor valleys and locally in the bottoms of minor drainage channels. Bauxite and other material were stripped from the major drainage areas. Bauxite was not formed in the major subsurface drainage routes between the nepheline syenite hills where erosion was taking place.

Bauxite and clay were removed from the place of formation and were carried toward the basin. The means of transportation probably ranged from landslides and soil creep to erosion and redeposition by stream action. This residual bauxite is referred to as the colluvial apron and has the same relationship to subsurface topographic features as the autochthonous deposits on weathered nepheline syenite farther up the slope".

One of the significant deposits occurs on the nepheline syenite of the Saline Dome (Fig.50). Nowadays the bauxite deposits surround igneous domes like gar-

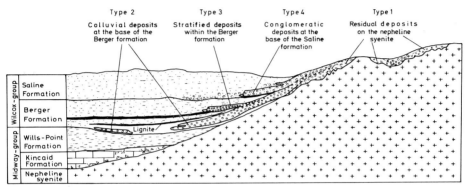

Fig.50. Schematic section illustrating dominant types of bauxite deposits on nepheline syenites of the Saline Dome, Arkansas, U.S.A. (After GORDON et al., 1958.)

lands. They are lenses, the diameter of which reaches several kilometres in many cases, with an average thickness of 3–4 m (maximum 10 m).

Laterally, the lenticular bodies pass into low-grade bauxites vertically zoned as follows (Fig.51):

    hardcap

    bauxite: zone of concretion

    bauxite: zone of leaching

    saprolite = underclay

    syenite.

Both the boundary of the fresh rock and the boundaries of the various zones are irregular and depend on local factors such as prebauxitic relief and drainage. Even in higher sections of the weathering profiles boulders with fresh syenite cores still occur.

Syenite. There is a whole group of alkali syenite (pulaskite, foyaite) with selective intense weathering, confined to varieties rich in nepheline rather than feldspars.

Saprolite (= underclay). There are greenish-grey kaolinite clays intercalated between syenite and bauxite, the lower portions of which may show syenite relic

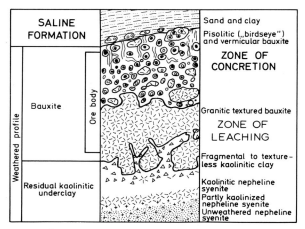

Fig.51. Bauxite profile on nepheline syenite of Arkansas, U.S.A., showing basal saprolite zone, lower bauxite horizon with relic textures and upper bauxite zone with pisolitic textures. (After GORDON et al., 1958).

textures. The thickness of this zone varies from 5–20 m but averages 10 m. These zones still contain boulders of fresh nepheline syenites with abrupt boundaries with the surrounding clay. Large feldspars, too, from coarse-grained pegmatites are preserved as relics in the weathering zone. The upper part of this zone consists of dense clays without relic textures. GORDON et al. (1958) believe the contact between the kaolinite zone and bauxite not to be the primary one but assume that the kaolinite clay originated from secondary resilicification.

Bauxites. The overlying bauxite is subdivided into two horizons – lower zone of leaching and upper zone of concretion – with distinct characteristics. The lower "granitic bauxite" defined as the "zone of leaching", is a soft, crumbly greenish-grey to pale rock, which partly preserved igneous rock textures but also developed typical "sponge ore" textures. The granular bauxite is mainly made of gibbsite. There is pseudomorphous gibbsite after feldspar, indicating direct transformation of feldspar into gibbsite. Cavities and veinlets are filled with gibbsite, opaque minerals, coarse-grained siderite and kaolinite.

Lower sections of the upper "zone of concretion", essentially originating from an influx of solution, are distinctly pisolitic, while there is rubbly texture in the upper part. Much of the rubbly appearance has been caused by the breaking apart and recementing of blocks of pisolitic bauxite, together with vermicular and vesicular forms and granitic-textured remnants.

Granitic-textured bauxite is dominant on steep slopes, whereas in more flatlying parts of the deposit pisolitic and rubbly concretionary ore is thicker.

Hard cap. The uppermost 30–60 cm of bauxite in many places is much harder or tougher than that below. The top of the bauxite also commonly contains more silica than the parts below. Pisolites and pebbles rest in a light to dark grey or brown cherty-looking matrix, which consists of halloysite and kaolinite.

Chemical composition. The various specimens from the zone of leaching and the zone of concretion are composed as shown in Table XXIV.

The enrichment of Al, Fe and Ti is highest in the middle profile sections, while silica is concentrated in the lower and upper parts (ADAMS and RICHARDSON, 1960). Several trace elements are enriched with aluminium (see Fig.78).

The bauxites formed during various stages, characterized by distinct paragenesis:

Neomineralization of the early phase: gibbsite formed from feldspar to a large extent directly. There are also primary iron oxides, such as goethite or hematite, and primary titanium minerals such as anatase. Whether there was primary kaolinite formation during weathering remains uncertain. Accessory minerals are ilmenite, sphene, fluorite and celsian.

During pisolite formation in the zone of concretion, boehmite replaced gibbsite to a certain extent. Locally co-existing boehmite and magnetite, probably also maghemite, indicate the secondary origin of those iron minerals. There are particularly hard pisolites which contain up to 70% Al_2O_3 in the "birdseye ore". KELLER (1964) suspects that the pisolites contain accessory corundum besides boehmite.

Secondary postbauxitic alterations are characterized by an influx of silica in a reducing environment under cover of younger sediments and ground water movements. Secondary silicification caused minerals of the kaolinite group to form in large amounts (see above) but there was also formation of chlorite and chamosite.

There are irregular *kaolinite veins* which enter the bauxite from underlying kaolinite clay. The veins reflect the fracture pattern of the syenitic source rocks. The jointing probably occurred postbauxitic giving rise to infiltration from above. Kaolinitization was not only confined to joints but kaolinite also filled cavities or replaced bauxite.

Resilicification of bauxite along fissures and joints resulted in a pseudo-fragmental texture, e.g., angular bauxite fragments resting in a predominantly kaolinite matrix.

Reduction of Fe^{3+} led mainly to the formation of siderite and pyrite, which impregnated the entire rock by filling cavities and veins or as concretions; the amounts of siderite formed may reach 30%. Often younger pyrite replaced siderite. Following uplift and denudation, ground water circulation in an oxidizing environment caused these rocks to decompose to a large extent, whereby limonitic crusts and sulphates formed.

TABLE XXIV

CHEMICAL ANALYSES[1] OF BAUXITE AND KAOLIN SPECIMENS FROM THE PRUDEN MINE, SALINE COUNTY, ARK., U.S.A. (after GORDON et al., 1958)

	Zone of concretions								
	1	2	3	4	5	6	7a	7b	8
Al_2O_3	45.6	61.2	56.7	61.2	60.7	60.9	59.4	61.7	57.6
SiO_2	24.8	3.7	6.9	1.8	3.0	6.9	3.8	1.1	5.8
Fe as Fe_2O_3	1.8	1.4	2.1	1.9	1.5	1.2	3.4	1.6	3.6
TiO_2	6.0	1.2	4.4	2.3	2.4	1.2	1.5	2.6	2.0
Ignition loss	21.2	32.2	29.1	32.4	32.2	29.6	31.2	32.6	30.4
Insoluble	0.6	0.3	0.8	0.4	0.2	0.2	0.7	0.4	0.6
FeO	0.2	0.2	0.2	0.2	1.4	0.2	0.2	0.2	0.2
	Zone of leaching								
	9	10	11	12	13a	13b	14	15	16
Al_2O_3	59.7	62.0	52.6	49.4	50.3	40.4	40.7	39.6	37.2
SiO_2	5.4	4.6	22.2	25.1	19.5	39.0	38.7	44.0	42.9
Fe as Fe_2O_3	1.1	0.9	0.8	1.1	3.0	2.5	2.4	1.1	0.8
TiO_2	3.7	0.4	0.8	1.4	2.6	1.9	2.5	0.6	1.9
Ignition loss	29.7	31.6	23.2	22.7	24.2	15.9	15.4	14.4	13.5
Insoluble	0.4	0.5	0.4	0.3	0.4	0.3	0.3	0.3	3.7
FeO	0.2	0.2	0.2	0.2	0.3	0.5	0.3	0.2	0.2

1. Gray partly kaolinized pisolites and rinds from near the upper surface of the bauxite. *2.* Red centers of pisolites associated with sample *1. 3.* Small block of bauxite with red pisolites enclosed in a drab matrix from the top of the deposit. *4.* Pisolitic bauxite block in red pisolitic bauxite ("birdseye ore") from the top of the deposit. *5.* Brown and white dense vesicular-appearing bauxite from the top of the deposit. *6.* White and pink scoriaceous bauxite from the upper 5 ft. of the deposit. *7a.* Orange-red bauxite interpenetrating drab vesicular matrix in a vermicular structure, from the upper part of the bauxite; orange material only. *7b.* Drab matrix from the same specimen. *8.* Hard tan ocherous, very porous granitic-textured bauxite ("sponge ore") from a block at the surface of the deposit. *9.* Greenish-gray to pale-tan crumbly granular (granitic-textured) bauxite from between two kaolinized nepheline syenite cores. *15.* Tan granular (granitic-textured) bauxite in place about 30 inches above sample *11. 11.* Grayish-tan soft granitic-textured bauxitic clay in the lower part of the deposit. *12.* Porous pale-gray clay in cut near the north edge of the deposit. *13a.* White granitic-textured bauxitic clay in contact with *13b. 13b.* Gray fragmental-appearing kaolinitic clay in an irregular body cutting granitic-textured and granular bauxite. *14.* Gray fragmental-appearing clay from near *13a. 15.* White granitic-textured kaolinitic clay adjacent to a hard nepheline syenite core. *16.* Light-bluish-gray kaolinized nepheline syenite adjacent to a core of fresher rock (15 and 16 underclay).

[1] Analyses by U.S. Bureau of Mines Field Laboratory, Little Rock, Ark., U.S.A.

Bauxite development similar to that of Arkansas is described from Pocos de Caldas in Brazil (HARDER, 1952) and Iles de Los in Guinea (LACROIX, 1913; MILLOT and BONIFAS, 1955; BONIFAS, 1959).

Indonesia and Malaysia

In Indonesia and Malaysia granites, metamorphics, sediments and particularly basic and intermediate eruptiva developed well-zoned laterites in Tertiary times (VAN BEMMELEN, 1941; WOLFENDEN, 1961, 1965; GRUBB, 1963). WOLFENDEN (1965) used the laterized andesite lavas in West Sarawak, Malaysia, to demonstrate the different epigenesis of "hill bauxites" and "swamps bauxites" governed by ground water. The hill bauxites rise approximately 30 m above the coastal alluvium, while the swamp bauxites occur below the Alluvium. Higher and lower pH and Eh-values above and below the water table, respectively, and different ground water circulation resulted in the transport of material in solution from hill bauxites to swamp bauxites. The weathering profiles are approximately 3 m thick. Between fresh-rock and bauxite there is a thick saprolite which formed below the water table. There are no relic textures preserved in the saprolite, but bauxites demonstrate the textures of source rocks very clearly. Gibbsite is pseudomorphous after plagioclase and goethite after ferromagnesium minerals. In hill bauxites the gibbsite content increased while silica and iron were removed. In swamp bauxites supplied with silica, quartz precipitated and silica reacted with gibbsite to kaolinite. There was reduction of Fe^{3+} followed by precipitation of amorphous gel or siderite matrix.

The mineral composition (in %) of hill bauxites and swamp bauxites is:

	gibbsite	kaolinite	goethite	quartz
hill bauxite	82.5	2.8	13.8	0.9
swamp bauxite	67.3	9.2	20.2	3.3

BAUXITES ON SEDIMENTS

General

Bauxite deposits of major economic importance occur on sedimentary rocks as two distinct groups: (1) intercalations of bauxite in clastic formations (Arkansas, Guiana, Weipa (Queensland); (2) as capping layers and pockets of bauxite in karst regions (U.S.A., Jamaica, Haiti, southern Europe, the Urals, China, Kashmir).

Historical

The first descriptions of bauxite deposits on karst were given by BERTHIER (1821) who referred to the occurrence at Les Baux in southern France. Several theories concerning their origin have since been developed.

(*1*) COQUAND (1871) believed that alumina and iron were transported in thermal solutions and flocculated on karst in southern France.

(*2*) DIEULAFAIT (1881) and LACROIX (1901) considered lateritic weathering products to be the original materials which were eroded from neighbouring igneous rocks and deposited on karst by rivers[1].

MALYAVKIN (1926) and later ARKHANGELSKY (1933) thought that iron and alumina originating from lateritic weathering at high levels were transported in colloidal suspensions, and flocculated from karst waters containing CO_2.

(*3*) Investigations of TUCAN (1912) and KISPATIC (1912) led to the conception of the autochthonous genesis of terra rossa (= red earth derived from residual clays of carbonate rocks) from which bauxite developed at a later stage.

DE LAPPARENT (1930) developed this theory further applying it to the southern European, in particular the French, bauxites. Many authors followed his principles of bauxite genesis in Mediterranean areas and extended these to deposits in Jamaica and Haiti.

Nowadays extensive geological and mineralogical investigations have covered many bauxite deposits and associated rocks. However, composition of primary strata, transport mechanism, place of sedimentation, diagenetic and post-diagenetic history differ at various localities. There are still many differences in opinion on the origin of strata-bound bauxites because of incomplete knowledge or generalization based on models of a particular deposit.

An allochthonous origin is assumed for coarse-grained deposits and those which are related to erosion of nearby laterites. In order to verify an autochthonous or allochthonous origin of fine-grained bauxites, priority must be given to the question of terra rossa genesis.

The following types of rocks may be distinguished: (*a*) coarse clastic bauxites; (*b*) terra rossa; (*c*) bauxites on clastic sediments; (*d*) bauxites on carbonate rocks; (*e*) bauxites on phosphate rocks.

Coarse clastic bauxites

Screes

Most bauxite deposits which rest on igneous or metamorphic rocks in areas of tropical or subtropical climate are of Cretaceous or Early Tertiary age. They are relics on old land surfaces, destroyed by erosion following uplift. The slopes are coated with boulders of bauxite material. Screes which surround plateau bauxites in India differ in texture and chemical composition depending on the speed of erosion. Typical are pisolitic high-quality bauxites rich in boehmite, which occur as block-like and very hard angular ores, formed along the original valley slopes.

---

[1] Several variations of the transport mechanism were considered by other scientists.

But there are also very hard spongy bauxites high in alumina and softer boulders originating from kaolinitic saprolite and the vesicular zone. The fine-grained detrital matrix of screes is inhomogeneous. The younger process of weathering in screes with good drainage systems produced high-grade ores through removal of iron and loss of SiO_2 from boulders rich in kaolinite. These screes are often mined because secondary loss of Si and Fe resulted in a high concentration of alumina. Another example of alumino laterites which rest on igneous rocks, the slopes of which are covered by screes, is that of the syenite domes in Arkansas. Similar to the bauxite screes of the plateau bauxites in India, coarse-grained screes 5–10 m thick occur on the slopes of the Saline dome in Arkansas interbedded with sediments of the Eocene Saline Formation (Fig.52). Downslope the screes die out, changing into sandy and calcareous clays. They form from the weathering of autochthonous bauxite in scree, slowly moving downwards. The scree consists of irregular and angular unsorted fragments to a maximum size of 1–1.50 m. The entire sediment is very heterogeneous with respect to mineralogy and texture. The detrital matrix is very inhomogeneous, consisting of bauxite, sandy clay and carbonates in varying proportions. The cement is also made of different materials, and gibbsite, kaolinite and carbonates occur in cavities. Gibbsite, kaolinite, siderite, leucoxen and pyrite are secondary mineral formations observed in the matrix.

The alumina content of these screes fluctuates widely, but is sufficiently high for mining. Only a few places show innumerable boulders of pisolitic bauxite.

Coarse clastic sorted sediments

A different type are the bedded deposits in the Berger Formation, which surround bauxite-covered syenites in Arkansas, U.S.A. The fluviatile coarse clastic bauxites are interbedded with sands, clays, bauxitic clays, sideritic strata and lignites of the Berger Formation (Fig.50). The deposits fill ancient channels, small swamps and depressions. They are lobate or tongue-shaped in plane, lenticular in cross-section. Beds of well-sorted bauxite gravel reach 3 m or more in thickness. Gradation and cross-stratification are common. They consist of multicoloured pisolites and bauxite pebbles, ranging from large rounded to subangular grains and granules about 2 mm in diameter to small pebbles and cobbles which reach 10 cm in length. Subspherical pisolites a few millimetres across are the most common. The pisolites are hard and consist mainly of submicroscopic gibbsite. They generally form layers 1–5 cm thick and are moderately well sorted. The cement consists of fine-grained gibbsite, kaolinite and granular siderite. Pisolites and matrix are cross-cut by younger kaolinite veins (Fig.52C, D).

The top section of these bauxite gravel deposits is impregnated with hard massive kaolinite. This may result from silicification caused by ground water movements. In places the top of the bauxite gravel beds contains enough granular siderite to constitute an indurated ferruginous capping.

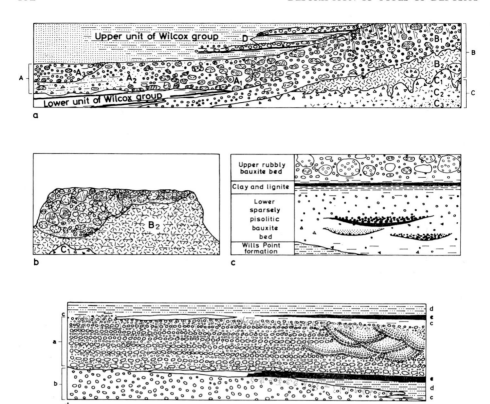

Fig.52. Detrital bauxites surrounding syenite domes in Arkansas, U.S.A.

a. Generalized cross-section of a type-4 deposit (Fig.50). The conglomerate bauxite and clay bed *A* lies at the base of the upper unit of the Wilcox Group below dark, carbonaceous clay interlaminated with sand. It overlies unconformably gray sandy to silty clay and lignite of the lower unit of the Wilcox Group. Some parts are mostly a coarse bauxite rubble (A1); others contain considerable clay and some sand (A2) and there are also channels filled with bauxite fragments and pisolites in a clay or bauxitic clay matrix (A3); bed *A* grades upslope into the upper part of type-1 deposit *B* of which pisolitic and vermicular bauxite of the zone of concretion (B1) and granitic-textured bauxite of the zone of leaching (B2) are shown. These overlie structureless or fragmental kaolinitic clay (C1) that grades downward through kaolinized nepheline syenite (C2) into fresh rock (C3). Small rubbly lenses *D* similar to the basal bed of the upper unit of the Wilcox Group *A* finger out into higher sediments of the Wilcox Group.

b. Bauxite boulder conglomerate of Saline age overlying granitic-textured bauxitic clay; B2 = zone of leaching; C1 = fragmental kaolinitic clay.

c. Diagrammatic sketch of channels filled with bauxite, sand and gravel in the Fletcher Mine, Saline Country, Ark., the vertical section represents about 13 m.

d. Generalized cross-section of a type-3 deposit (Fig.50). Stratified and crossbedded bauxite *a* of a type-3 deposit unconformably overlies pisolitic bauxite *b* of a type-2 deposit. Bauxite in both deposits is overlain by kaolinitic clay *c*. Encroaching sediments of the Berger Formation include gray silty to sandy clay *d* and lignite *e* that grades upslope into a lens of sideritic rock (shown by light tone). Kaolinitic veining beneath lignite and kaolinitic clay in the upper part of both bauxite deposits is indicated by dotted pattern. (After GORDON et al., 1958.)

Further examples of coarse-grained, well-sorted bauxites are many ores in the Urals and European Mediterranean areas: deposits of Gant, Hungary; Var, France (pp.72,159).

Terra rossa

The classical terra rossa investigations were carried out by Tucan (1912) and Kispatic (1912) on karst of Dalmatia on Lower Cretaceous limestones. They postulated progressive changes from autochthonous residual clays via terra rossa to bauxite. It is accepted today that bauxite may form from terra rossa and that lateral transitions from terra rossa to bauxite exist.

However, modern terminology distinguishes between soils originating in situ from limestones and allochthonous material reworked several times resting on limestones.

Mückenhausen (1962) classified:

terra calcis

terra rossa	terra fusca
typical terra rossa	typical terra fusca
"verbraunte" terra rossa	
"rendzinierte" terra rossa	"rendzinierte" terra fusca

Weinmann (1964) investigated the soils of the Kefallinia Island, Jonian Islands. These soils have formed in situ on karst of Mesozoic and Eocene carbonate facies under Mediterranean climatic conditions[1]. He found that close relations existed not only between soil formation and primary rock, climate and vegetation but also in relief. Formation of terra rossa started in Tertiary times, and the process is still continuing below 800 m as "harmonic soil". Terra rossa and terra fusca are contemporaneous final products resulting from development at different morphological levels.

The mineralogical and chemical composition given in Tables XXV and XXVI characterize autochthonous profiles of terra rossa and terra fusca on limestones.

Apparently those soils lack free hydroxides of aluminium, or they occur in only minor amounts. However, red earths on Tertiary limestones of Jamaica rich in alumina are described by Richard (1963). They contain up to 10% gibbsite in concretions.

The terra rossa on Jurassic limestones of the Velebit Mountains in Yugoslavia (southwestern Dinaric Alps) are believed to be autochthonous by Marić (1967). He assumes that there was no supply of clastics at the high altitudes at

---

[1] With high annual rainfall but dry summers.

TABLE XXV

CHEMICAL COMPOSITION OF TERRA CALCIS, SIZE <0.002 mm (after WEINMANN, 1964)

Profile Nr.	Horizon Nr.	SiO_2 (%)	R_2O_3 (%)	Fe_2O_3 (%)	Al_2O_3 (%)	$SiO_2/$ R_2O_3	$SiO_2/$ Fe_2O_3	$SiO_2/$ Al_2O_3	$Al_2O_3/$ Fe_2O_3	Free Fe_2O_3
20	A 48	39.94	39.61	11.52	28.09	1.91	9.22	2.41	3.82	7.84
Terra rossa	B_1 49	41.74	41.03	11.32	29.71	1.92	9.81	2.39	4.11	7.96
	B_2 50	41.99	42.01	11.44	30.57	1.88	9.76	2.33	4.19	8.12
29	A 73	43.93	36.74	10.99	25.75	2.28	10.63	2.72	3.67	3.78
Terra fusca	B_1 74	43.00	36.47	10.97	25.50	2.25	10.42	2.86	3.64	4.44
	B_2 75	42.73	39.07	10.72	28.35	2.06	10.60	2.56	4.14	4.45

% Al_2O_3 calculated by difference R_2O_3—Fe_2O_3.

TABLE XXVI

EXCHANGEABLE CATIONS AND CLAY-MINERAL CONTENT OF TERRA CALCIS, SIZE <0.002 mm (after WEINMANN, 1964)

Profile No.	Horizon No.	Exchangeable cations (mequiv./100g)	Clay-mineral content
20	A 48	35.08	illite > vermiculite > kaolinite, quartz > montmorillonite > attapulgite, chlorite
Terra rossa	B_1 49	36.50	illite > vermiculite > kaolinite, quartz, mont-morillonite > attapulgite, chlorite
	B_2 50	39.42	illite > vermiculite > kaolinite, quartz, mont-morillonite > attapulgite, chlorite
29	A 73	45.58	illite > vermiculite > kaolinite, quartz, mont-morillonite > attapulgite
Terra fusca	B_1 74	42.11	illite > vermiculite > kaolinite, quartz, mont-morillonite > attapulgite
	B_2 75	43.92	illite > vermiculite > kaolinite, quartz > attapulgite

which karst formation was in progress. Six million tons of terra rossa formed from residual clays of limestone accordingly. There are gibbsite and nordstrandite in nodules and oolites (Al_2O_3 up to 57%) besides montmorillonite-illite and kaolinite minerals. In other parts of the Dinaric Alps (MARIĆ, 1964) redeposition of terra rossa and neomineralization were observed. This mountain range extends northwest-southeast. In the east it is made of highly metamorphic Palaeozoic sediments associated with ophiolites and granitic bodies. In the southwesterly direction these rocks are covered with Permotriassic, with clastics and evaporites, Triassic, Jurassic, Cretaceous and Palaeogene carbonate facies in that order. Dacites, andesites and basalts penetrate this sedimentary Mesozoic belt.

In successive phases, Jurassic and Cretaceous limestones and dolomites developed a pronounced karst topography, nowadays covered by terra rossa. Hematite, goethite, kaolinite-chlorite and montmorillonite minerals are new phases which occur together with residual minerals such as quartz, mica, feldspars and stable heavy minerals. In addition there are widespread traces of specific relic minerals from: (*1*) metamorphics (epidot, kyanite, andalusite, staurolite, amphibol); (*2*) Permotriassic (anhydrite); (*3*) bauxites (boehmite, diaspore, corundum).

In certain areas a sandy clastic facies of terra rossa with secondary silicifications named "saldame" developed, which was deposited by a river or lake. Therefore, MARIĆ (1964) assumes that terra rossa originates from both carbonate rocks and, in the eastern belt region, igneous, metamorphic and sedimentary rocks.

Removal of silica in terra rossa is caused by: either autochthonous soil formation on karst, which leads to relative enrichment of silica in the upper and relative concentration of alumina and iron in the lower part of the vertical profile (FILIPOVSKI and CIRIC, 1963) or frequent redeposition from high-level areas to the littoral regions of the Adriatic Sea in the west.

The amount of SiO_2 carried into the Adriatic Sea at present is very high. The Pantan Spring near Divulje supplies the Adriatic Sea with 300 l/sec in late fall and 2,000–3,000 l/sec in springtime, at which time it contains 1,400 mg/l Si. The annual amount disposed of by this spring alone is 100 tons of SiO_2.

The lateral transport of terra rossa accompanied by removal of silica is proven herewith. The following scheme modified from MARIĆ (1964) summarizes the features.

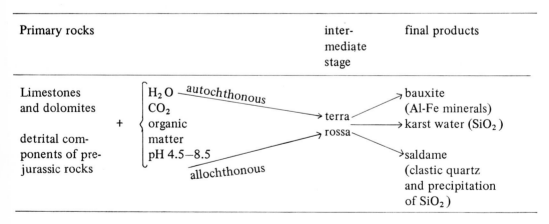

Primary rocks		intermediate stage	final products

Bauxites on clastic sediments

Huge bauxite deposits recorded in Tertiary clastic sediments formed east of the Guyana shield. The world's biggest bauxite deposits occur on the Cape York peninsula, Australia. From the Deccan peninsula bauxite deposits on Precambrian phyllites have been described.

Coastal plain east of the Guyana shield

The Precambrian Guyana shield is made of metamorphics and sediments intruded by granites and dolerites in the Proterozoic. The plateaus of the old land surface are covered with Al laterites from Cretaceous to Early Tertiary age. According to CHOUBERT (1965) there are 400 million tons of bauxite (42% Al_2O_3) in the Kaw Mountains (French Guyana), and DOEVE and GROENEVELD MEIJER (1963) give the figure of 300–400 million tons of bauxite (45% Al_2O_3) in the Bakhuis Mountains and on other equivalent plains.

In the northerly pitching coastal plain of Guyana younger sedimentation commenced in Cretaceous times (Takutu Formation). Further to the south, however, sedimentation started in the Eocene. Late Tertiary strata rest unconformably on Eocene successions. There is also a marked decrease in thickness of transgressive Quaternary coastal sediments from Guyana in a southerly direction. Only the basin of Berbice is filled with Quaternary sediments of considerable thickness (up to 2,500 m).

In Tertiary times rivers meandered, and thus reduced the speed of material transport from the highland into the sea.

Clastic sedimentary successions above deeply weathered basement consist of irregular lenticular layers of interbedded gravel, sands, kaolinitic clays and lignites.

Age	Pollen zones	Surinam		Guyana Bauxite Belt
Recent		DEMERARA		recent sediment in river valleys
Pleistocene	G2	COROPINA	Lelydorp / Para	
Pliocene	G1?			MACKENZIE Formation
Miocene	F	COESEWYNE	Upper / Lower	MONTGOMERY Formation
Oligocene	E		?	?
Eocene	D / C	Bauxite	?	Bauxite
Paleocene	B2	ONVERDACHT	Upper / Lower	MOMBAKA Formation
Maastrichtian	B1 / A	?		?
Precambrian		Basement		Basement

Fig.53. Bauxite occurrences in the Tertiary of Surinam and Guyana (compiled from MONTAGNE, 1964; and VAN DER HAMMEN and WYMSTRA, 1964).

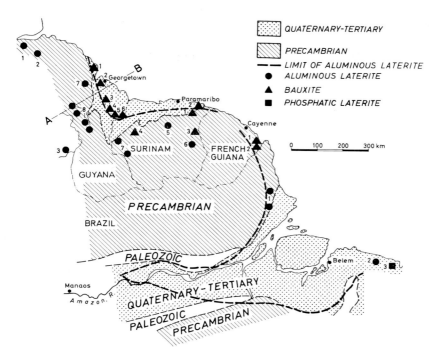

Fig.54. Distribution of Al laterites on basement and corresponding bauxites in the Tertiary coastal plain. (After BLEACKLEY, 1960; and PATTERSON, 1967.)
Venezuela:
1 = Upata region; *2* = Nuria region; *3* = Gran Sabana region.
Guyana:
1 = Pomeroon district; *2* = Esse quibo district; *3.* = Mackanzie district; *4* = Ituni district;.
5 = Kwakwani district; *6* = Canje district; *7* = Blue Mountains; *8* = Pakaraima Mountains
Surinam:
1 = Surinam River region, Paranam, Onverdacht, Rorac, Para River deposits; *2* = Moengo and Rickanau deposits; *3* = Nassau Mountains district; *4* = Bakhuis Mountains district; *5* = Brown Mountains; *6* = Lely Mountains; *7* = Wana Wiero Hills.
French Guiana:
1 = Roura deposits; *2* = Kaw Mountains deposits.
Brazil:
1 = Amapa region; *2* = Piria River region; *3* = Ilha da Trauira.

In this fan-shaped delta close to the highland predominantly coarse clastic facies with gravel, quartz sand, pellets and nodules of iron oxide formed, while kaolinite suspensions precipitated in swamps, lakes and lagoons near the coast. There are several disconformities within the sedimentary series.

The palynological study of VAN DER HAMMEN and WYMSTRA (1964) seems to point to the conclusion that there is only one bauxite horizon in Guyana, which must have formed between pollen zones B and E, that is to say most probably somewhere between Early Eocene and Early Oligocene (Fig.53).

The depositional environment of the kaolinite clays may have been swamps in an embayment at a river mouth with hills covered by laterite, a coastal strip with cliffs and islands. The economically important bauxite deposits nowadays rest partly on basement and partly on Palaeocene sediments. The thickness of which decreases towards the coast. Plateau bauxites and bauxites on sediments are probably of the same age (Fig.54, 55).

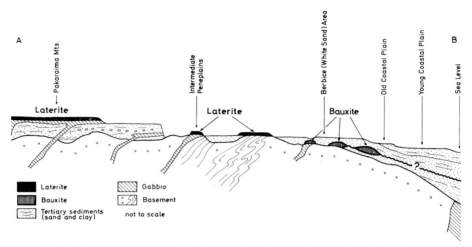

Fig.55. Structural geology of laterites and bauxites in the coastal plain of Guyana; profile *AB* in Fig.54. (After BLEACKLEY, 1960.)

Autochthonous bauxite formation on the sedimentary land surface and deposition of detrital material from the plateau overlapped in Kwakwani, Guyana (MOSES and MICHELL, 1963). Therefore, the final stage of bauxitization was accomplished only in places of long exposure. Kaolinite clays remained preserved as bauxite equivalents in areas of rapid sedimentation. HARDEN and BATESON (1963) and DOEVE and GROENEVELD MEIJER (1963) studied the relationship of bauxite quality to drainage related to relief of capping and underlying strata. ALEVA (1965) investigated the chemistry and genesis of these bauxites in Onverdacht, Surinam. The gibbsite-bauxite layer at this locality is interbedded with clay, sands and lignites, the thickness of which increases to the north, whereas the strata die out in a southerly direction.

The bauxite bodies are irregularly shaped and vary in thickness from 0–17 m but average 6 m. Frequently stratification is still observable, and enrichments of heavy mineral placers may be detected, indicating a primary rock of clastic nature, perhaps arkose.

Authigenic textures. The consistency of bauxite ranges from hard rock (90% of the bauxite mined requires blasting) to plastic clay. The textures of neomineralization distinguished are: concretionary bauxite, massive bauxite, cellular bauxite, banded bauxite (with relic textures of stratification, caused by arkose-clay rhythm and crossbedding layers of heavy minerals; Plate II, 22), breccia-like bauxite and globular bauxite concretions of 3–30 cm in diameter.

The various bauxite types are confined to distinct horizons of the bauxite body. The concretionary type belongs to the upper and middle part, while cellular and banded types occur in the bottom part of the deposit. The breccia-like bauxite appears exclusively near or at the surface of the beds, while the iron-rich type forms pockets or inclusions in solid layers of bauxite low in iron.

The massive type does not form continuous horizons but develops irregular bodies up to 1 m in diameter amidst the other types or in the underlying kaolinite clays.

Boundaries of bauxite deposits. The mean level of the upper edge is mainly horizontal. A surface of extremely thin bauxites rests lowest. There is no relation between the lower boundary of the bauxite deposit and its top. Typical for the floor is a widely branching, irregular slope pattern with steep-sided hills of quite limited extent formed by the underlying kaolinite clay. The surface of the bauxite stratum reflects the character of an old land surface. Gullies and rivers cut into the bauxite deposit leaving behind a slightly domed plateau. Development of drainage systems occurred during and after bauxite formation.

Within the top part of the bauxite stratum occur isolated caps rich in iron. These are relics of the originally 1–3 m thick crusts which probably later selectively weathered to bauxite low in iron with breccia-like texture (Fig.56). Top sections both rich and low in iron are traversed by vein systems filled with silica-rich kaolinitic bauxite or with metahalloysite. These are the resilificated and iron-leached equivalents of the original iron-rich cap, formed on the land surface through action of water and vegetation or after being covered by younger sediments. These successive stages may lead to complete removal of iron and resilicification of bauxite.

In most cases the base of the bauxite stratum forms a sharp but very irregular contact with the underlying kaolinite clays. Within 2.5 m, without any apparent change in the bauxite surface, the average ore thickness of 6–8 m may increase to 15–17 m. Bauxite dykes (cellular type) and pipes (massive type) with vertical or steeply inclined walls have been found, penetrating for unknown distances into the underlying kaolin.

In the lower part of the bauxites, particularly in the vicinity of the boundary with the underlying kaolin, banded bauxite is typical. It may be rich in layers of coarse-grained heavy minerals (rutile, anatase, tourmaline, staurolite, sphene). Thin interbedded lenses of kaolin may also occur as well as clay seams which enter the bauxite floor. These phenomena may be interpreted as selective bauxitization of specific materials.

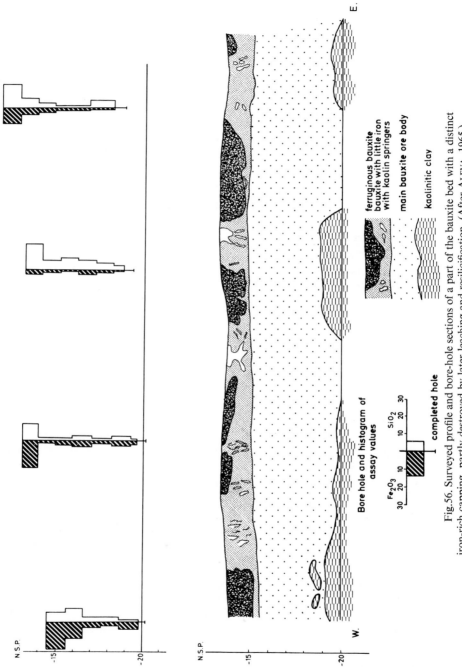

Fig.56. Surveyed profile and bore-hole sections of a part of the bauxite bed with a distinct iron-rich capping, partly destroyed by later leaching and resilicification. (After ALEVA, 1965.)

Chemical composition. The chemical aspects are shown in the histograms *A–D* in Fig.57. Histogram *A* shows the distribution of bauxite thickness in bore holes investigated. Histogram *B* gives the high Fe_2O_3 and SiO_2 values in positions close to the top of the profiles. Approximately 6% iron is assumed to be present in the original sediment; secondary removal has reduced this value. The high silica content results from resilicification. The original mean value of SiO_2 for bauxite was 2.5%. This is likely to be 1/20 of the prebauxitic sediment. Histogram *C* demonstrates that there is no relationship between SiO_2 content and the height of the uppermost part above the base of the bauxite deposit. However, there is a strict correlation between the amount of Fe_2O_3 and the thickness of the bauxite deposit. This suggests a migration of iron from all parts of the bauxite stratum to its topmost section. Besides this the iron-rich top sections are always somewhat higher than portions low in iron. There are two possible interpretations, both of which may be true: bauxites beneath low-level valleys possibly contained less iron, or the soft parts of the bauxite surface eroded readily while the hard crusts of iron oxides and hydroxides formed inselbergs.

The correlation between SiO_2 content and bauxite thickness is the reverse of that for iron, as shown in histogram *D*. However, the high SiO_2 value obtained for thin bauxite strata results from the secondary infiltration of SiO_2 from overlying sediments in the upper sections. The thickness of the underlying bauxite bears no influence on the high SiO_2 content of the topmost portion. The iron content of the upper iron crust increases with thickening of the bauxite layer because more iron becomes available. The original correlation between bauxite thickness and enrichment of iron in the capping crust was altered later through mobilization of iron at the bauxite surface.

We arrive at the following theories on genesis:

(*1*) The original differences in bauxite source material and the underlying clay may be explained by:

(*a*) the sharp boundary between bauxite and the underlying kaolinite clay;

(*b*) differently sized heavy minerals in bauxite and kaolin;

(*c*) the frequent occurrence of banded bauxite layers and kaolinitic lenses in the lower part of the bauxite stratum.

According to ALEVA (1965), the original sediment of the bauxite in Surinam probably was of an arkosic or silty nature with coarser grains and higher porosity (30%) than the underlying kaolin. There is no decrease in bauxitization towards the floor.

BLEACKLEY (1960), however, points out that the present ore bodies in Guyana throughout the deposit show evidence of having evolved mainly as gels from colloidal solutions, e.g., shrinkage cracks, brecciated and colloform textures and rhythmic precipitation of material of different composition in cryptocrystalline and amorphous form, being the normal textures of typical bauxite.

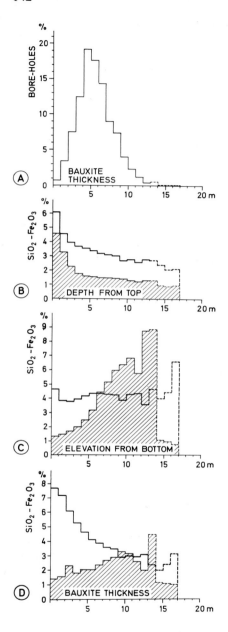

Fig.57. Histograms representing the results of a statistical investigation of chemical analyses. For explanation, see text. SiO_2 histogram in heavy lines, Fe_2O_3 histogram shaded. (After ALEVA, 1965.)

(2) Bauxitization occurred over a long period. Pure gibbsite formed in all areas with good drainage and sufficient porosity. Under optimum drainage conditions, up to 15% boehmite was formed.

(3) The transport of iron occurred in a vertical direction, and there was secondary enrichment in upper sections with increasing thickness of bauxite.

(4) The high SiO_2 content in the upper sections is independent of the thickness of the bauxite stratum. Circulating ground water rich in SiO_2 reacted with the bauxite surface under the younger sediments. The iron caps dissolved, and kaolin minerals formed in vein systems.

(5) Bauxitization continued after subsidence and resulted in the formation of cellular bauxite by solution. Little is known about the paragenetic development in these bauxites.

Cape York peninsula, Australia

Huge bauxite deposits which rest on a succession of clastic sediments, probably Early Tertiary in age, are described by EVANS (1959) in the flat, low lying and relatively heavily timbered terrain near Weipa, Australia; LOUGHNAN and BAYLISS (1961) noted continuous weathering on arkose sandstones since Tertiary times there. The investigations are of special interest not only because of the high SiO_2 content (90% SiO_2 and 4% Al_2O_3) of the primary rock but also because there is ample scope to observe the relationship of chemical composition and mineralogy to morphology, level of water table and pH values. The profile embraces three units (Fig.58): (3) concretionary zone (with pisolitic textures); (2) zone of fluctuating water table; (1) original sediment.

The original sediment (1) is a kaolinitic arkose sandstone. Zone 2 is approximately 6 m thick, of deep red colour and earthy to massive in texture. It essentially consists of quartz and kaolinite. Quartz is mostly angular, and solution effects are rare. Hematite and goethite form typical irregular concretions and are enriched together with kaolinite approximately 1 m above the base of the zone. At higher levels kaolinite disappears, being replaced by gibbsite and in the uppermost part, partly by boehmite.

The concretionary zone, zone 3, is up to 5 m thick and distinctly separated from zone 2. Small amounts of quartz occur throughout this part of the profile. Pisolites, frequently cemented, predominate and contain gibbsite, hematite and/or goethite. The pisolites range from 1–20 mm in diameter. Usually the matrix is formed by Al hydroxides and kaolinite. The amount of boehmite is highest in this zone, which was formed by dehydration of gibbsite in the opinion of LOUGHNAN and BAYLISS (1961). The pH value is about 5.5 and rises to 7 below the water table. Dark soil, approximately 1 m thick, covers the bauxites.

Genesis. Hematite and/or goethite enrichments in zone 2 may be caused by:

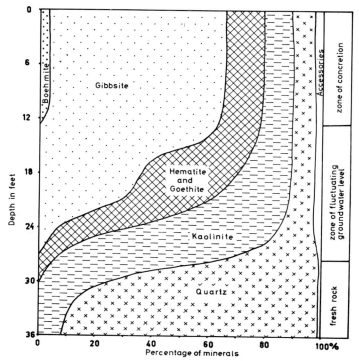

Fig.58. Mineral distribution in bauxite profiles on kaolinitic sandstone from Weipa, Qld., Australia. (After LOUGHNAN and BAYLISS, 1961.)

annual fluctuation of the water table; persistence of dissolved iron in a reducing environment below the water table; rapid oxidation to insoluble Fe^{3+} compounds through periodic exposures in a warm and oxidizing environment; high atmospheric temperatures which favour the accumulation of leaves and other organic matter at the surface but eliminate the reducing power of the ground water.

The most important feature is the migration of iron in an upward direction and the precipitation of stable iron compounds in the zone of oxidation. V^{5+} and Fe^{3+} (similar ionic radii) form constant proportions in the upper section of the profile. There is an increase of Al-values towards the direction of a lower relief.

No detailed investigation has yet been published describing the age and mechanism of Al-accumulation. So far we have no explanation for such huge accumulations of alumina, but lateral influx of aluminous solutions could be regarded as a possible source.

Munghyr area, Kharagpur hills, India

ZIA-UL HASAN (1966) describes laterites with bauxite horizons from the

Munghyr area, India, which forms caps on Precambrian shales. In the present areas the source-rock–bauxite sequence is exposed in many places on the upper hill slopes, revealing the characteristic products, resulting from the alterations that the source rock has undergone.

Profile descriptions. Shales with quartz, feldspar and mica altered either to gibbsite–goethite–anatase bauxites or to goethite–gibbsite–kaolinite laterites rich in iron. Traces of boehmite and diaspore occur in some bauxites.

From Precambrian slates and phyllites of the Kharagpur hills, India, ALWAR (1962) describes similar caps of lateritic weathered rocks which pass laterally into both massive and pisolitic bauxites.

Data on profiles of bauxitic weathering on clastic sediments are concerned mainly with: (*1*) facies pattern (catena); (*2*) lateral penetration of solutions (which probably are of major importance); (*3*) diagenetic and post-diagenetic phases of paragenesis; (*4*) geochemical processes.

The examples given demonstrate that "latosols" rich in alumina form from clastic sediments containing high proportions of feldspars and layer silicates. Relic textures of the original rock in these vertically zoned profiles are superimposed by textures of diagenetic mineral formations. The mineralogical and chemical composition in specific weathering zones is analogous to weathering zones in profiles on igneous rocks. Also facies changes from laterite to bauxite are probably the equivalent of facies pattern (catena) on igneous rocks.

Bauxites on carbonate rocks

Bauxite deposits on karst vary greatly in size and developed throughout the earth's history, occurring all over the world. The main ages of formation are plotted in Fig.6.

HARRASSOWITZ (1926) distinguished "Kalkbauxite" from "Silikatbauxite" (formed on igneous rocks). VADAZ (1951) proved that bauxites on Cretaceous limestones in Hungary contain allochthonous material. He, therefore, replaced the word "Kalkbauxite" with the neutral term "Karst-Bauxite" in order to express the relationship to bedrocks but to avoid any statement concerning the origin of the material.

Nowadays the term "Karst-Bauxite" is widely used because more and more bauxite investigations indicate allochthonous origin of source materials, and more frequent discoveries of clastic intercalations between underlying limestones and bauxite are made.

The three foremost questions of genesis deal with: (*1*) distributive province, source material and transport mechanism; (2) diagenesis; (3) post-diagenetic processes.

Distributive province, source materials, transport mechanism

The distributive province of these bauxite deposits embraces areas of the stable or unstable shelf region. The sedimentary troughs involved a range from several hundred metres to several hundred kilometres in length. The bauxite thickness may vary from 1 m to a maximum of 50–60 m. The average thickness is 3–8 m.

There are three theories concerning the source material for karst bauxites:

(1) Bauxitization of residual clays in situ: Mediterranean countries: DE LAPPARANT (1930); DUBOUL-RAXAVET and PÉRINET (1960); Jamaica and Haiti: HARTMANN (1955); HILL (1955); BUTTERLIN (1958); HOSE (1963); CLARKE (1966); AHMAD et al. (1966); SINCLAIR (1967).

(2) Parautochthonous transport of residual clays from high-level areas into basins: DENIZOT (1934); BONTE (1965); KOMLOSSY (1967).

(3) Allochthonous cover on karst: Mediterranean countries: VADAZ (1951); HABERFELLNER (1951); ERHART (1956); ROCH (1956, 1967); BARDOSSY (1958); VALETON (1965, 1966); DEMANGERON (1965); NICOLAS et al. (1968); Jamaica and Haiti: ZANS (1954, 1959); EYLES (1958).

To prove autochthonous origin for bauxitized residual clays on karst is difficult and successful only if the original textures of the carbonate rocks were preserved in bauxites. Such results have not yet been reported. Parautochthonous redeposition of residual clays of carbonate sediments in axial depressions is likely for portions of the prebauxitic sediment. The percentage of parautochthonous material generally remains unknown because of lack of positive identification.

An allochthonous cover of karst is apparent in many cases from geological environment or mineralogical criteria. Essentially there are two groups of source material:

(1) Lateritized or fresh material of variable grain sizes deposited mainly by rivers in terrestrial basins or under marine conditions close to the coast.

(2) Fresh eolian volcanic tuff transported.

The following examples demonstrate the different possibilities in distributive provinces, source materials and transport mechanisms.

Gulf coastal plain, U.S.A.

In the southern part of the coastal plain of the U.S.A. some bauxite deposits rest either on clastic sediments or on carbonate rocks. They are a transitional type between bauxites in clastic successions and those on carbonate facies. The most important factors governing bauxite deposit formation, are (OVERSTREET, 1964): (1) the pre-existence of lateritized igneous complexes; (2) fossil land surfaces; (3) protection of bauxites and kaolin clays from erosion.

Inner and outer bauxite belts formed around the igneous complex of the Blue Ridge and Piedmont Province on a Late Cretaceous to Early Tertiary land surface, slightly tilted towards the Early Tertiary Gulf of Mexico and the Mississippi embayment (Fig.59). The deposits on the northwestern flank of the inner belt rest on karst of Palaeozoic carbonate strata of the Valley and Ridge provinces, the Valley Floor or Harrisburg peneplain (BRIDGE, 1950). The surface was probably not a featureless plain but included low hills and ridges. If so, erosion and stream deposition were active processes.

The bauxite deposits on the southeastern flank, however, are located on Late Cretaceous sands and clays. Formations of Middle and Late Eocene age and Oligocene age overlie the bauxites in the Irvington district.

The bauxites of the outer belt (BRIDGE, 1950) occur on the eastern Gulf Coastal Plain and are interbedded with clastic successions between the Midway and Wilcox Group (Palaeocene–Eocene interval; Fig.60).

In Mississippi (Fig.60, *1–3*), bauxites and associated kaolinitic clays rest at the top of the Potters Creek clay which grades downwards through white micaceous clay into dark grey micaceous clay making up most of the formation. Probably the clay which originally constituted the upper part of the formation in the bauxite areas was different from most of the formation, representing a shoreward facies of the Potters Creek clay. The marine equivalent is highly montmorillonitic. On the southeastern flank in Alabama (*5*) and in Georgia (*6–7*) bauxites and kaolinite are restricted to the Nanatalia Formation of Early Eocene age (Fig.61) with underlying karst of the marine Palaeocene Clayton Formation. Between the Eufalia district (*5*) and the Winston Nocubee Kemper district (*3*) marine shallow water sediments of Palaeocene and Early Eocene age occur. No evidence of the shoreline remains, and if bauxite ever was present along the margin of the bay, it has been removed by erosion. The preservation of bauxite deposits from erosion is an important factor in their present distribution.

The age was determined by fossil plants found in kaolinite clays associated with the bauxites. The period of bauxite formation in the two areas was short, beginning not earlier than Late Paleocene and ending in the Early Eocene.

In the Valley and Ridge province the shape of the bauxite deposits depends on karst topography related to cleavage and joint patterns. The ore lenses may range from one to several hundred metres in length, and most are thicker than they are wide. Dolinas from the Wilcox prospect, Smith prospect and Indian mount district (DUNLAP et al., 1965) are coated with residual clays rich in chert from the underlying strata. Lens-shaped irregular successions of sands, lignite and kaolinite clays attributed to the Wilcox Group, cover the deposit. There is evidence of postbauxitic slumps by small slickensides, particularly in the kaolin, and by intermixing of displaced masses of different types of bauxite and blocks of pisolitic bauxite in a kaolinitic matrix.

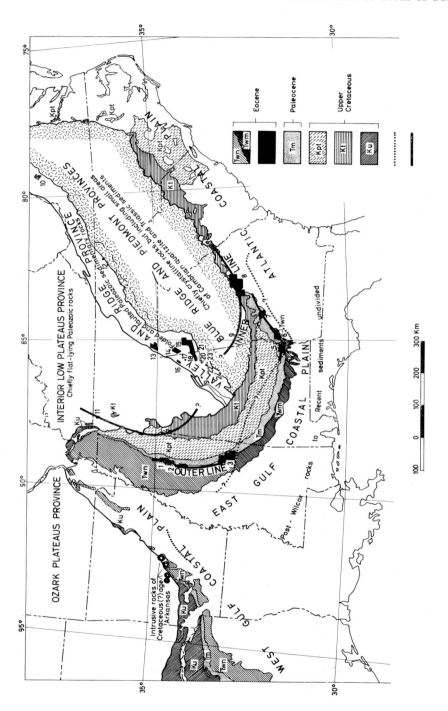

Age	Group	West Tennessee Indian Mount. district	Northwest Alabama Margerum districts	Mississippi All districts	Eastern Alabama Eufaula districts	Western Georgia Springvale and Andersonville distr.	Central Georgia	
							Irwinton district	Warm Springs district
EOCENE	Jackson				Jackson Group	Jackson Group	Jackson Group	
EOCENE	Claiborne				Claiborne Group	Claiborne Group	Claiborne Group	
EOCENE	Wilcox	Clay of probable Paleocene and Eocene age	Clay of probable Paleocene and Eocene age	Wilcox Formation Fearn Springs Sand Member	Bashi Marl Member of Hatchetigbee Formation	Bashi Marl Member of Hatchetigbee Format.		
EOCENE	Wilcox	Clay of probable Paleocene and Eocene age	Clay of probable Paleocene and Eocene age	Wilcox Formation Fearn Springs Sand Member	Tuscahoma Sand	Tuscahoma Formation		
EOCENE	Wilcox	Clay of probable Paleocene and Eocene age	Clay of probable Paleocene and Eocene age	Wilcox Formation Fearn Springs Sand Member	Nanafalia Formation	Nanafalia Formation		
PALEOCENE	Midway			Potters Creek Clay	Clayton Formation	Clayton Formation		
PALEOCENE	Midway			Clayton Format.	Clayton Formation	Clayton Formation		
LATE CRETACEOUS	Selma			PrairieBluffChalk and Owl Creek Fm.	Providence Sand	Providence Sand	Upper Cretaceous rocks, undifferenti-ated	Upper Cretaceous rocks, undifferenti-ated
LATE CRETACEOUS	Selma			Ripley Formation	Ripley Formation	Ripley Formation	Upper Cretaceous rocks, undifferenti-ated	Upper Cretaceous rocks, undifferenti-ated
LATE CRETACEOUS	Tusca-loosa	Sand and gravel probably equi-valent to Gordo Formation	Gordon Formation					
MISSISSIPPIAN			Cypress Sandstone					
MISSISSIPPIAN			Gasper Formation					
MISSISSIPPIAN			Bethel Sandstone					
MISSISSIPPIAN			Ste.Genevieve Limestone equivalent					
MISSISSIPPIAN		Warsaw Limestone	Warsaw Limestone					

———— Bauxite

Fig.60. Correlation chart of formations exposed in bauxite districts of the coastal plain and outliers of coastal plain rocks. (After OVERSTREET, 1964)

The kaolinite clays which fill karst depressions and dolinas pass via low-grade bauxite into high-grade bauxite from the margin towards the centre. The bauxites have both clay-like non-pisolitic and hard pisolitic textures, the latter being composed of nearly pure gibbsite.

Fig.59. Bauxite districts and areas of the southeastern United States and their relationship to certain formations and major physiographic divisions. *Ku* = Late Cretaceous rocks undivided; *Kt* = Tuscaloosa Group or Formation undivided; *Kpt* = Post-Tuscaloosa Cretaceous strata; *Tm* = Midway Group undivided; *Tw* = Wilcox Group undivided; *Twn* = non-marine phase; *Twm* = marine phase.
Dots: approximate position of the early Wilcox shoreline; heavy line: inner and outer lines of deposits in the coastal plain province; black areas: bauxite.
Districts and areas: *1*. Tippah-Benton, Miss.; *2*. Pontotoc district, Miss.; *3*. Winston-Noxubee-Kemper district, Miss.; *4*. Margerum district, Ala.; *5*. Eufala district, Ala.; *6*. Springvale district, Ga.; *7*. Andersonville district, Ga.; Areas adjacent to and between Andersonville and Springvale districts, Ga.; *8*. Irwinton district, Ga.; *9*. Warm Springs district, Ga.; *10*. Spottswood district, Va.; *11*. Indian Mound district, Tenn.; *12*. Elizabethton district, Tenn.; *13*. Chattanooga district, Tenn.; *14*. Summerville area, Ga.; *15*. Hermitage, Cave springs, and Bobo areas, Ga.; *16*. Fort Payne area, Ala.; *17*. Congo area, Ala.; *18*. Rock Run and Goshen Valley areas, Ala.; *19*. Ashville area, Ala.; *20*. Jacksonville area, Ala.; *21*. Nances Creek area, Ala.; *22*. Anniston area, Ala.; *23*. De Armanville area, Ala.; *24*. Talladega area, Ala. (After OVERSTREET, 1964.)

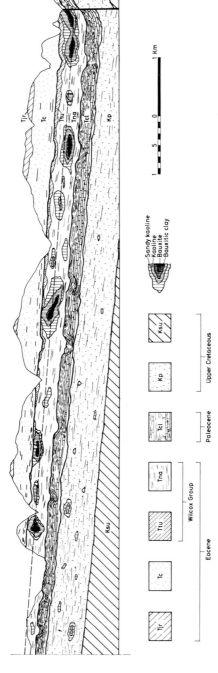

Fig.61. Relationship of stratigraphic units and ore bodies in the Andersonville bauxite district, Ga., U.S.A. Bauxites form lentilles with the highest Al content in the central parts. *Tjr* = clay unit of the Jackson Group, includes residuum from limestone of Oligocene age; *Tc* = sand of the Claiborne Group; *Ttu* = Tuscahoma Formation; *Tna* = Nanafalia Formation; *Tcl* = Midway Group, Clayton Formation, limestone with karst topography; *Kp* = Providence Sand; *Ksu* = sedimentary rock, undifferentiated. (After ZAPP, 1965.)

This type of bauxite deposit unambiguously developed from sandy-clayey terrestrial sediments redeposited on karst following fluvial transport from high-level areas. The straightforward relationship between thickness of the clay and quality of intercalated bauxite lenses indicates autochthonous bauxite genesis following sedimentation on a fossil land surface.

Jamaica and Haiti

ZANS (1954) studied the deposits of Jamaica in detail and also examined karst hydrology (1951). Since the Miocene uplift the very thick Tertiary yellow and white limestone formation has undergone strong karstification, particularly in the upper sections. The water table, led by numerous karst flows from the entire limestone formation, is governed by the impervious floor made of andesites, tuff and tuffitic clay of Cretaceous and Early Tertiary age. In the Central Range the structural height of the older igneous rocks is higher than the surface of the limestones. The springs of both surficial and karst streams are located in volcanic rocks, the outcroppings of which, in the Central Range, developed lateritic soils which supplied the karst area with huge amounts of lateritic weathering material of andesite and tuff through erosion during the wet seasons (Fig.62).

Following transport over long distances the fine fractions were deposited in seasonal lakes in large karst depressions. In Jamaica transitional stages range from terra rossa rich in silica and residual clays of limestones to bauxites containing very little silica. At high levels with good drainage, predominantly gibbsitic bauxites rich in alumina formed, while terra rossa rich in kaolinite developed at lower levels in topographic depressions with periodical stagnation of ground water movement. The Jamaican bauxites are textureless (concretionary particles are very rare), and very seldom lithified, thus forming dominantly earthy masses. The lateritic soil of Haiti is similar in composition to soil deposits in the limestone valleys of the Sierra de Bahoruco in the Dominican Republic and on the limestone upland of Jamaica. All these deposits rest on Tertiary limestone and are probably closely related in origin (GOLDICH and BERGQUIST, 1948).

European Mediterranean areas

Bauxites on karst are very widespread in southern Europe (Fig.63). During the final stage of the development of the Alpine geosyncline cycles of uplift were followed by denudation and marine sedimentation. Intermittently peneplains formed and lateritic weathering took place. In contrast to the American deposits which formed around more stable high-level areas, connections of source area and place of sedimentation are frequently disrupted here. Moreover distributive provinces themselves were dislocated by young tectonic movements and partly eroded.

The oldest, small bauxite intercalations occur in the Permian of Turkey. In the

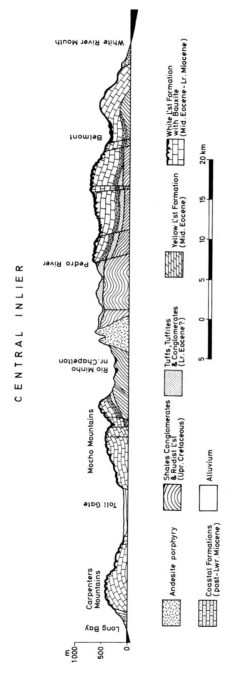

Fig.62. Section across the central part of Jamaica, showing the structural relations of the bauxite-bearing white limestone formation and the older strata. Vertical exaggeration, 7 ×. (After Zans, 1954.)

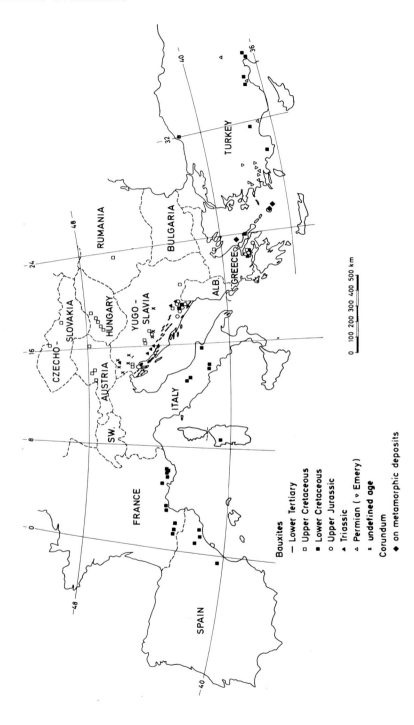

Fig.63. Distribution of bauxites in southern Europe.

Triassic of Yugoslavia, at Smartno ob Paki, Slovenia (Anisian-Ladinian age) and at Gracac, Croatia (Ladinian-Karnian age), coarse pisolitic bauxites occur which are cross-bedded and may be interpreted unambiguously as fluvially redeposited older pisolitic bauxites.

The oldest bauxites of economic interest occur in the Late Jurassic of Yugoslavia and Greece. Extensive deposits of Cretaceous age are widespread, particularly in the Early Cretaceous of Spain, France, Italy, Austria, Yugoslavia, Hungary, Greece and Turkey. The youngest bauxites spread over regions of Italy, Yugoslavia, Greece and Turkey, are of the Eocene age.

Most Mediterranean bauxites rest directly on carbonate rocks without intercalations of clastic sediments. Frequently only relics of the original distributive provinces are preserved. Lack of fossils or other means of correlation render palaeogeographic reconstructions difficult, ambiguous or even impossible. In addition, different types of bauxite deposits result from a variety of post-diagenetic processes. The complexity of their late history renders the interpretation of distributive provinces and original sediments difficult.

The longest time interval with repeated bauxite intercalations in Turkey ranges from Permian to Eocene and, in Yugoslavia in the Dinaric Alps, from Ladinian to Lutetian time. A bauxite horizon, not very typical, still occurs in the Miocene. Fig.64 illustrates the features schematically.

Basal contact. In France a belt of Early Cretaceous bauxite deposits stretches from Haute Var in the east to Ariège in the western foreland of the Pyrenees. The bauxites form flat lenses 10–30 km in length, several hundreds of metres in width and averaging 3–8 m in thickness, or fill dolinas in karst of Jurassic or Early Cretaceous strata. COMBES and REY (1963) mapped intercalations of bauxites in the limestone reefs of "Urgonfacies" of Ariège. Bauxite developed continuously on the Central Belt region, but ore lenses occur in several stratigraphic horizons in the upper part of Early Cretaceous strata in areas with interfingering marine sediments. The unconformity between bauxites and carbonate strata of different ages provides a sharp contact. The relief of the carbonate facies is governed by: (*1*) petrology; (*2*) joint patterns; and (*3*) level above the water table.

Karstification progresses more rapidly in Middle Jurassic dolomites than in Late Jurassic limestones and is intensified by dense joint patterns. Similar features of karst topography may be observed in Yugoslavia, Hungary and Greece (MACK, 1964).

Regionally one may subdivide in the same manner: (*a*) areas with dolinas on undulating karst surfaces (Plate VI, 23); (*b*) closed basins which contain bigger bauxite deposits on relatively flat karst surfaces.

Late Cretaceous bauxites in Greece reflect a relationship between regional uplift and karst topography of the underlying limestone (NIA, 1968).

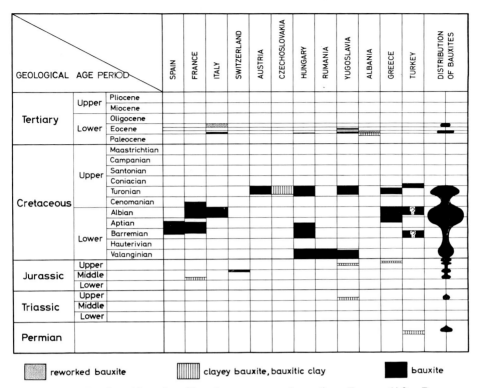

Fig.64. Stratigraphic order of bauxite occurrences in southern Europe. (After BARDOSSY, 1963.)

Karst topography developed discontinuously: in France at least four stages of karstification may be distinguished (VALETON, 1965): (*1*) the main stage of karstification which is prebauxitic; (*2*) continuation of karstification during sedimentation of bauxite source material; (*3*) local karstification which followed diagenesis (immediately); and (*4*) youngest karstification governed by modern topography.

The main stages of karstification prior to and during sedimentation have been observed at many localities. The transported detrital material accumulated in karst depressions. The floor of the bauxite bodies reflects the prebauxitic and synsedimentary karst topography. Particularly thick lateritic sediments bauxitized later resulted from the development of synsedimentary graben structures in Hungary (Fig.65; SZANTNER and SZABÓ, 1962; SZABÓ, 1964).

Immediate post-diagenetic local karstification may be observed near Mazaugues in southern France, where limestones unconformably overly lithified but collapsed and shattered bauxites.

Post-bauxitic karstification on dislocated limestones, governed by modern

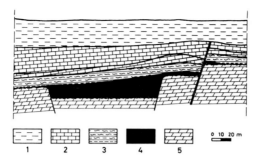

Fig.65. Synsedimentary movements gave rise to the formation of particularly thick bauxite profiles in the graben zone, Hungary.

Profile across the bauxite area of Nagytárkány Austrian and Pyrenean orogenic phases: *1*. Sarmatian clay with limestone detritus; *2*. Middle Eocene limestone; *3*. Early Eocene limestone; *4*. Early Eocene marl, clayey marl, carboniferous clay; *5*. Cretaceous bauxite. (After SZANTNER and SZABÓ, 1962.)

topography, are very common in southern Europe. They may be recognized from plugs perpendicular not to the former but to the Recent land surface (VADAZ, 1951; COMBES, 1965; PAPASTIAMATIOU, 1965).

Clastic intercalations between bedrock and bauxites may vary in type. In France stratified limestone breccias may be found in many dolinas below bauxites. Characea of lacustrine environment or marine fossils of the Urgonian occasionally occur in the matrix (ROCH, 1956, 1966; BONTE, 1965; NICOLAS and ESTERLE, 1965). Interbedded clay with lignitic material has been described by NICOLAS (1968) from Mazaugues in southern France. In Rumania (LUCCA, 1966) breccias, conglomerates and sandy sediments occur between bauxites and the very pure limestone bedrock. The clastic sediments are reworked detrital material from the igneous rock zone of the Monts Apuseni. In these bauxites there are also intercalations of sandy and clayey sediments with fossils, including plants, poorly preserved.

Very seldom a progressive change from limestones to overlying bauxitized clastic sediments has been described. PAPASTIAMATIOU (1965) observed in the Late Cretaceous bauxite (third horizon) of Kafénèdés (Helicon) in Greece a progressive change from limestones to volcanic tuff with amphiboles and feldspars. The tuff in turn passes into pisolitic bauxites with a calcareous matrix.

Upper boundary. In general, the originally upper boundaries of the bauxite deposits which were horizontal during sedimentation are now slightly undulating because of differential compaction. Occasionally, fine distinct stratification may still be observed in the uppermost part of the bauxite deposit.

Obvious progressive changes to terrestrial lignitic clayey roof sediments may be seen at times. This may indicate facies changes, disconformities or continuous sedimentation. Usually there was very little reworking by marine transgression which closely followed bauxitization. However, in slowly subsiding areas with long periods of terrestrial exposure, soil development, formation of lignites or fresh-water sediments, considerable reworking of bauxites occurred in places to form finely stratified bauxite sediments or bauxite breccias above the bauxite bodies.

Shape of bauxite bodies. The bauxite deposits are shaped mainly by the contours of overlying and underlying strata. Nowadays they are either dolina fillings or lenses several hundreds of metres in width and 10–15 km long. In Var (south-eastern France), the bauxite seam was originally (Fig.66) a broad, flat, talus fan around the Maure-Esterelle Massif (VALETON, 1966). The thickness of the bauxite

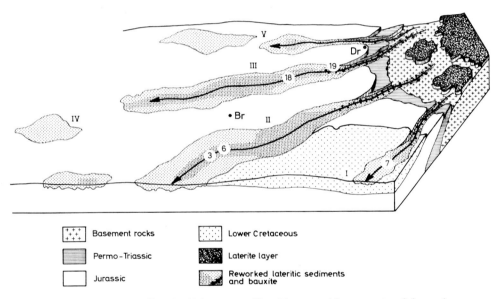

Basement rocks

Permo-Triassic

Jurassic

Lower Cretaceous

Laterite layer

Reworked lateritic sediments and bauxite

Fig.66. Maure-Esterelle Massif in eastern Var, France, with remnants of formerly extensive fans of reworked lateritic material. The remnants are confined to basins protected from erosion following postbauxitic structural displacement. *Dr* = Draguignan; *Br* = Brignoles. I. syncline of Le Revest (profile *7*); II. syncline of Pelican-Engardin-Mazaugues-St. Baume (profiles *3*, *6*); III. syncline of Pas de Recou-Combecave-Le Val (profiles *19*, *18*); IV. syncline of Pourcieux; V. syncline of Villecroze-Fox-Amphoux. (After VALETON, 1966.)

ranges from 2 to 8 m. Several long finger-shaped, closed basins radially orientated around the massif resulted from more recent tectonic movements and denudation. Bauxites throughout the Mediterranean countries form similar flat Alluvial fans. The relation between source area and distributive province may be seen particularly well in Greece. The lateritic cover of the ophiolites in the sub-Pellagonic zone eroded, and colloids rich in iron, silica and alumina were deposited on karst in the nearby Parnassus-Kiona zone (Fig.67).

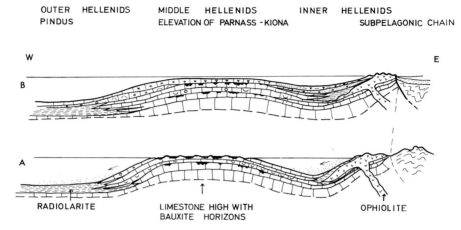

Fig.67. Relationship of source area (ophiolitic sub-Pellagonic zone) and bauxite formation on the heights of carbonate rocks of Parnassus-Kiona in Greece. A. In the upper section of Early Cretaceous and during Middle Cretaceous. B. During Early Senonian. (After CELET, 1962.)

The nature of the *source material* depends on the detrital products available at high-level areas of denudation which probably varies from locality to locality.

For most bauxite deposits it is difficult to assess grain sizes and the mineralogy of transported materials, which would indicate the transport mechanism, as relic textures are poorly preserved in general. In Greece in the sub-Pellagonic zone allochthonous origin of nickeliferous bauxite on karst of Jurassic limestones is obvious because of the neighbouring lateritized ophiolites. The bauxites of the Parnassus-Kiona zone too are characterized by higher values of nickel and chromium.

Clastic materials from high-level areas with lateritized basalts or keratophyrs have been identified (WIPPERN, 1965) as the source of both Permian diasporites and Cretaceous bauxites in Turkey. Lateritized sands and tuffs deposited on karst of Early Permian limestone changed into bauxite in situ. Near Bolkardag there is a facies change from diasporites into quartz conglomerates with a hematitic-diasporitic matrix towards the massif. In the Early Cretaceous bauxites of Akseki in Turkey, Wippern found relics of fresh feldspars and contours of disintegrated feldspars. Sourse rocks are considered to be lateritic weathering products of albitized ophiolithes of neighbouring high-level areas.

The bauxites surrounding the Maure-Esterelle Massif, contain clastic relic minerals such as quartz, tourmaline and zircon. In the surroundings of Draguignan in close proximity to the massif, the bauxite fillings of dolinas contain clastic angular quartz (maximum diameter 1–3 mm). This coarse-grained quartz does not originate from residual clays of carbonate sediments; its only source are volcanic rocks of the Maure-Esterelle Massif (after ROCH and DEICHA, 1966). The high nickel content (0.5–2.0% NiO) also proves the allochthonous origin of these bauxites (VALETON, 1966).

Radioactive minerals also indicate the allochthonous origin of bauxites. Radioactivity in bauxites of Gant in Hungary varies greatly. The intensity is twice as high as in the Valence granites. Layers with enrichments of carnotite relics $K_2(UO_2)_2V_2O_5 \cdot 3 H_2O$ in the bauxites of Unterlaussa in Austria further prove an allochthonous origin. There is a 1 cm wide rim of bleached bauxite surrounding carnotite grains (KÖHLER, 1955).

Transport mechanism and sedimentary environment. The bauxites of Unterlaussa in Austria are stratified and change laterally into limestone conglomerates (RUTTNER and WOLETZ, 1956). There are intercalated lenses of clayey and lignitic sediments in the lower part of the bauxites of Gant in Hungary. KISS and VÖRÖS (1965) believe in the selective precipitation of more aluminous material or alumina silica compounds from suspensions depending on pH. Fluvial transport is indicated by cross-bedded pisolitic layers in the upper part of these bauxites[1].

Opinions differ concerning the size of the transported grains and the transport mechanism for bauxites in France. CALLIÈRE and POBEGUIN (1965) believe in transport in solutions. On the basis of textural investigations, VALETON (1966) assumes that periodic mud flows formed a talus fan surrounding the Esterelle Massif. In close proximity to this massif coarse and poorly sorted components were deposited (Plate VII, 27), while with increasing distance precipitation from suspension became the dominant feature. NICOLAS (1968) observed graded bedding near Mazaugues. There are also intercalations of large marine brecciated limestone lenses resulting from subaqueous slides. NICOLAS (1968) describes the environment of bauxite sedimentation in the western part of the Var as marine-lacustrine or lacustrine because of horizons with marine fossils (brachiopods, pelicipods, gastropods) in the bauxites of Brignoles in France. The transport-media were mud flows or subaqueous slides. Nicolas assumes that bauxites including pisolitic textures existed in the source area prior to erosion. In their opinion the bauxites are clastics with minor diagenetic alterations.

Other bauxites were deposited in a marine environment as proven by the occurrence of marine fossils in the Late Cretaceous of Greece (NIA, 1968), Unterlaussa 2, Austria (HABERFELLNER, 1951), and the Middle Eocene of Dalmatia with marine molluscs, corals, echinoids, bryozoans and terrestrial plants: SAKAČ (1966).

---

[1] Limnic gastropods like *Pyrgulifera* cf. *pichleri, Melania heberti, Strophomostella cretacea* are found in Late Cretaceous bauxites of Hungary (BARDOSSY, 1957; BARNABAS, 1961).

The Urals

In the northern Urals (BUCHINSKY, 1963), conglomerates 2–3 m thick with limestone pebbles cemented by bauxite occur between Palaeozoic limestone bedrock and bauxite. In the southern Urals bauxite breccias cemented by calcite occupy these stratigraphic positions.

Many bauxites in the Urals (e.g., Krasnaya Shapochka) consist of psammitic or conglomeratic strata in which no or poor sizing can be detected. The material of these bauxites appears to have been transported by temporary mud or turbidity flows. This is also indicated by sorted lamination or graded bedding (the grain size decreases gradually from bottom to top in each rhythm of stratification). Such flows entered karst caves where turbid water settled, with sedimentation beginning with the coarser fractions. The layers of fine bauxite mud exhibit dessication fissures filled with psammitic bauxite.

The transported material was deposited as talus on the slopes and as coluvium at the bottom of the slopes—near-contact type of BUCHINSKY (1963)—or else as lake sediment in intermittent or perennial impounded bodies. Usually no conglomerates mark the change from terrestrial and terrestrial-fresh-water bauxites to overlying marine sediments. Bauxite-bearing lowlands apparently have been flooded by the sea, forming gulfs or lagoons.

Pacific Islands

Finally we refer to the red earths rich in gibbsite on coral reefs of the Pacific Islands. Very similar $Al_2O_3/Fe_2O_3/TiO_2$ ratios in both volcanic rocks and bauxites on karst are given by KALUGIN (1967). He assumes sedimentation of tuff on carbonate facies, from which terra rossa and finally bauxite developed.

Most recent observations suggest that:

(*1*) Source materials may comply mineralogically with bauxites and laterites, clays, arkose, tuff or similar rocks.

(*2*) Grain sizes may range from boulders of allochthonous origin to colloids.

(*3*) Sedimentation may occur from turbulent periodical rivers as fanglomerates or from smooth turbidity currents in lakes or lagoons or as subaqueous slides.

(*4*) The sedimentary environment may vary from terrestrial (periodical rivers or lakes) to marine (lagoons).

(*5*) The bauxites of southern France, as postulated by H. Erhart show a distinct sedimentary succession. Denudation of laterites on neighbouring high-level areas first removed the top section rich in Fe, Si and Al and this was followed in turn by removal of the central part rich in Fe and Al and the bottom section rich in silica and alumina. A reverse profile developed accordingly in the source rock of bauxite. (An alternative interpretation will be suggested in the section on diagenesis.)

(*6*) The change from bauxite to overburden normally is abrupt. The sequence is: bauxite-lignite or lignitic clay–fresh-water sediment–marine sediment.

Diagenesis

Before entering into the discussion of bauxite diagenesis we consider:

(*1*) The nature of source materials may vary from arkose to lateritic sediments or autochton residual clay of different compositions. Grain sizes range from boulders to colloids.

(*2*) Diagenesis always takes place in terrestrial basins at different levels above the water table.

(*3*) Fracture patterns of the bedrock, orientation and size of dolinas and karst depressions govern ground water movements.

(*4*) Climate and vegetation determine environment in the downward direction of the profile.

(*5*) Environment during sedimentation of the overburden may influence late diagenetic mineral formations of the ore.

In many cases classification of source material remains doubtful due to lack of textural features. However, there are unambiguous examples of source materials accumulated unaffected by diagenesis as well as bauxites with strong autochthonous diagenetic alterations.

Clastic rocks hardly altered by diagenesis are known from the stratified pisolitic bauxite sediments (Carboniferous) of the Urals, from the Karnian bauxite of Dalmatia, from Early Cretaceous bauxites close to the Massif Central of Maure (Plate VII, 27) in southeastern France and from the Late Cretaceous bauxites of Gant in Hungary.

At other localities karstbauxites developed in situ from lateritic material by autochthonous diagenesis. Relic clay minerals may still be recognized where bauxites pass into stratified clays. In southwestern France montmorillonite and illite clays laterally turn into bauxites. In southeastern France (NICOLAS, 1968) and in Greece (NIA, 1968), upper sections of the bauxite bodies, only slightly altered by diagenesis, still show stratification. The kaolinitic source minerals differ widely in lattice order at these locations. Unambiguous examples of autochthonous diagenesis are bauxites in Jamaica, Hungary, France, Greece, the U.S.A. and the Urals. Bauxite genesis is characterized by both removal of silica and dehydration of Al- and Fe-hydrates in situ (VADAZ, 1951).

Present knowledge distinguishes the following diagenetic processes:

(*1*) Changes in chemical and mineralogical composition: (*a*) desilicification–transformation of clay minerals into Al-hydroxide hydrates; (*b*) dehydration–transformation of Al-hydroxides into hydrated Al-oxides and Al-oxides; (*c*) partial removal of iron; (*d*) basal enrichments of certain trace elements.

(2) Changes in textures: (a) development of vesicular textures, pisolitic textures or breccia-like textures; (b) lithification.

Changes in textures and chemical and mineralogical compositions are contemporaneous.

The argument is frequently expressed that there is a relationship between neomineralization of gibbsite-boehmite-diaspore and the age of the deposit. Indeed most gibbsite bauxites on karst are of Tertiary age, such as the deposits of Jamaica, Haiti and southern Europe. Exceptions are the Early Tertiary diaspore bauxites of Kashmir. There are lateral facies changes from gibbsite bauxite to boehmite bauxite, to diaspore bauxite in the Cretaceous of southern Europe. Moreover, in the Palaeozoic bauxites of North America and the Urals boehmite and diaspore occur together.

Depending on the local environment, but independently of geological age, diagenesis results in:

(1) kaolinite–gibbsite facies

(2) kaolinite–gibbsite–boehmite facies

(3) kaolinite–boehmite facies

(4) kaolinite–boehmite–diaspore facies

(5) boehmite–diaspore facies.

The examples in following sections illustrate possible courses of diagenesis.

Jamaica

The mineralogical composition of bauxites rich in gibbsite have rarely been investigated so far, and little is known about facies and environment during diagenesis. In bauxites of Jamaica the kaolinite–gibbsite–boehmite facies is best known.

The dominant clay mineral is kaolinite in places where the deposit is rich in silica. There are also traces of chlorite. Some montmorillonite occurs in clays of dolinas from lower levels with logged ground water (AHMAD et al., 1966). Much silica has been removed from these clays in the central part of the dolinas due to drainage joints of the karst (Fig.68). Neomineralization of aluminium hydroxides occurred. In terrestrial oxidizing environment gibbsite dominates boehmite at higher levels of the deposits, but it is subordinate in lower terrain. The very high iron content (11–23% Fe_2O_3) results from present hematite and goethite, as small ferruginous concretions or pellets, especially in the upper part of the deposits. MnO_2 and P_2O_5, too, are enriched in the concretions.

The usual chemical composition of Caribbean terra rossa bauxite is given in Table XXVII.

In permeability and porosity, which approach 50%, this type of spongy bauxite closely resembles loess. This is indicated by the degree of saturation of the available pore space, ranging from 60–98% in samples collected towards the

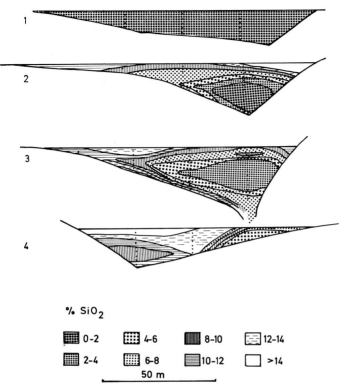

% SiO₂

▦ 0-2   ▦ 4-6   ▥ 8-10   ▤ 12-14

▦ 2-4   ▦ 6-8   ▤ 10-12   ▢ >14

50 m

Fig.68. Silica values are lowest in central parts of dolinas where drainage intensity is highest; Jamaica. (After Zans, 1954.)

TABLE XXVII

CHEMICAL COMPOSITION (IN %) OF CARIBBEAN TERRA ROSSA BAUXITES

	Jamaica Higher Plateau (Hose, 1950)	Haiti Rochelois Plateau (Goldich and Bergquist, 1948)	Dominican Rep. Sierra de Bahoruco (Goldich and Bergquist, 1947)
SiO₂	0.4– 3.5	2.4– 5.3	1.55– 5.17
TiO₂	2.4– 2.6	2.3– 3.1	2.50– 2.75
Al₂O₃	46.4–50.3	42.6–49.4	46.25–48.53
P₂O₅	0.3– 2.8	0.3– 0.8	0.13– 0.26
Fe₂O₃	17.5–22.8	20.8–23.5	19.43–20.61
MnO₂	not determined	0.4– 0.7	0.13– 0.56
MgO	not determined	0.04–0.18	not determined
CaO	0.1– 1.2	trace–0.18	not determined
H₂O (below 110 °C)	not determined	1.8– 2.6	0.73– 1.53
Loss on ignition	26.0–27.8	20.1–25.8	23.43–26.55

end of the dry seasons. During the wet seasons a tremendous amount of water is absorbed.

Southern Europe

The Upper Cretaceous bauxites of Hungary show transitional stages from clays, bauxitic clays via clayey bauxites to high-grade bauxites. According to BARDOSSY (1958) the high-grade bauxites which are products of the most advanced stage of diagenesis, form a northeast–southwest axis (Fig.69). Lateral diagenetic

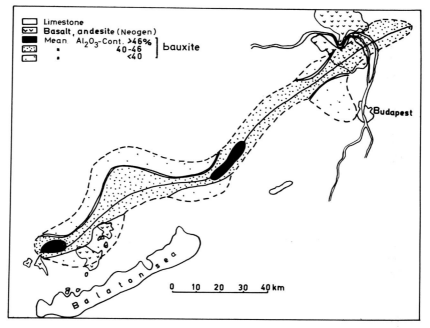

Fig.69. Relationship of geography and mean Al_2O_3 content of bauxite occurrences in Hungary. The central bauxites are highest in alumina with similar distribution of Fe_2O_3 and TiO_2, but silica is lowest along the central axis. (After BARDOSSY, 1958.)

differentiation of clay to bauxite may also be noted in profiles investigated by SZANTNER and SZABÓ (1962). There is high-grade bauxite of considerable thickness in structurally low positions whereas clays or clayey bauxites rest on the belts. The size of the basin and thickness of sediments are related to the quality of bauxites.

Apparently the lateral facies change is caused by desilicification in situ. The same lateral facies change from clays to bauxites may also be detected in bauxites of France, in the forelands of the Pyrenees, and in areas surrounding the Maure-Esterelle Massif (VALETON, 1965, 1966). There is desilicification, and three mineral facies (as in Hungary) co-exist: gibbsite facies, boehmite facies and diaspore facies.

Little is known about the gibbsite facies of the Alpilles and the Languedoc. The boehmite–hematite facies is the rule for most bauxites of the Var. Lateritic detrital material varying in grain size was deposited in various depressions surrounding areas of denudation (see above). Different levels of the basins above the water table caused variations in ground water movements, resulting in different intensities of diagenesis.

In the northern troughs (Fig.66) argillites rich in kaolinite formed. In the depressions of Le Val and Mazaugues, where the most advanced stages of diagenesis are found, thin bauxite in structurally high positions contains more SiO_2, and autochthonous desilicification is clearly related to relief (Fig.70). Vertically the bauxites form a main section 2–3 m thick, dominantly ferrallitic in character and capped by a siallitic layer a few decimetres thick. Both layers developed contemporaneously but their boundary is irregular (Plate VI, 24).

During the initial stages of diagenesis pisolites ranging from 0.2 to 2.0 mm in diameter formed in the ferrallitic zone; if their size is below 0.2 mm these are frequently termed oolites in the literature. They are of concretionary origin and are named pisolites or oolites in this context. The centers consist of either relic fragments of lateritic iron compounds or ferrallitic gels. The shells grew from the rhythmic precipitation of iron and alumina forming fine layers of predominantly boehmite or hematite. The submicroscopic intergrowth of boehmite and hematite crystallites indicates crystallization of both minerals from a gel (Plate VIII, 29, 33) rich in Fe and Al. During pisolite formation, the material was ductile causing pisolites to be distorted uniformly (Plate IX, 35) and giving preferred orientation over a larger area by flow. The pisolites are cemented by a submicroscopic matrix of hematite and boehmite. The uniform sequence of crystallization in neighbouring pisolites, the uniform deformation pattern of the pisolites and lateral transition to non-pisolitic bauxite indicate growth in situ of the pisolitic textures.

The upper siallite zone is rich in silica, but there is little iron. With the naked eye textures are hard to discern in these rocks, but there are microtextures of kaolinite–boehmite concretions embedded in a kaolinite matrix. The uniform textures suggest contemporaneous paragenetic development in ferrallite and siallite zones. The bauxite surface is penetrated by fossil roots, indicating vegetation during bauxite diagenesis (Plate IX, 38, 39). At very low pH values in the upper section, resulting from vegetation, silica reacted with gels rich in alumina to form kaolinite. Iron compounds dissolved, and Fe migrated from this horizon. The mobilization of iron extended to the lower ferrallite zone with repeated dissolution and precipitation of Fe. A pseudobreccia texture in the bauxite developed from the selective disintegration of hematite. Angular or fringed particles varying in size occur in a light-coloured matrix. In other places iron accumulated in fossil water tables and precipitated as a goethite crust on the denser and iron-enriched pseudobrecciated parts (Plate VIII, 29–31). This iron mobilization occurred during the final stages of diagenesis and frequently yielded extensive dark inclusions in the white bauxite matrix.

The pattern of iron removal reflects the drainage system during the final stages of diagenesis. It is related to the joint pattern of the karst bedrock (Plate VII, 25; X, 40).

Besides the boehmite facies there are bauxites in France (Aude, Ariège) showing all transitions from boehmite bauxites via boehmite–diaspore bauxites to diaspore bauxites. Initial precipitation in the matrix and pisolites resulted in submicroscopic boehmite or boehmite–hematite masses. Diaspore is secondary and coarse grained and grew over the primary textures in the pisolites (Plate IX, 36).

The three facies are characterized by the dominating alumina minerals gibbsite, boehmite and diaspore in each case. In all three facies hematite is the primary iron mineral. While secondary goethite formed in the gibbsite and boehmite facies, either iron migrated from the diaspore facies or magnetite or siderite (Plate IX, 34) formed at a later stage (refer to post-diagenetic paragenesis).

The clay minerals in the boehmite facies are well-crystallized kaolinites, but the lattice order is relatively poor in the diaspore facies where other layer silicates, such as Al-chlorite, illite and montmorillonite, also occur (CAILLÈRE et al., 1962).

The proportion of amorphous material in the three facies varies. The boehmite facies is completely crystalline, while the diaspore facies includes a larger amount of amorphous material. Contraction through loss of water was highest in the diaspore facies which is traversed by numerous veinlets of diaspore and kaolinite. The same minerals formed at various stages of crystallization in the matrix, in pisolites and concretions and in dehydration cracks and are pseudomorphous.

Autochthonous desilicification and dehydration with diaspore formation is best known from the detailed investigations of bauxites in Greece by NIA (1968). There, too, relations existed between highlands of the sub-Pellagonic zone of ophiolites supplying lateritic weathering products and a karst relief of the Parnassus-Kiona zone. The lateritic weathering products were trapped in karst depressions during Late Cretaceous times. Marine gastropods and lacustrine fauna and flora indicate precipitation of a suspension rich in iron, silica and alumina in littoral regions. Fine stratification of the original sediment is indicated in the upper sections of the bauxites rich in clay. Conversion to bauxite took place in a terrestrial oxidizing environment. Desilicification and dehydration processes were studied in a close network of specimens in several dolinas. Results showed distinct relationships between: (*1*) relief and drainage; and (*2*) relief and distribution of Si, Al and Fe.

Increasing drainage resulted in decreasing silica content and increasing alumina values. Iron has been removed from both the upper sections of the profiles and parts with optimum drainage. Titanium became enriched primarily with iron and secondarily with alumina.

Selective dissolution and removal of the components resulted in textural alteration and mineralogical composition. Both porosity and density were highest in areas with optimum drainage (Fig.71, 72), i.e., highest rate of removal developed a maximum of inner surface, where minerals of highest density formed.

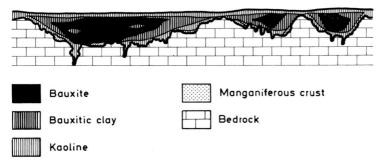

	Bauxite		Manganiferous crust
	Bauxitic clay		Bedrock
	Kaoline		

Fig.70. Schematic section through bauxites of Mazaugues. (After VALETON, 1965.)

The main minerals are kaolinite, boehmite, diaspore, goethite and hematite. Because kaolinite is the only silica-bearing mineral, the kaolinite distribution coincides with the SiO_2 content. Kaolinite mainly occurs in top sections and in zones of poor drainage in contact with underlying strata. In areas with optimum drainage diaspore was formed from primary boehmite during the final stages of diagenesis. Laterally and vertically pure diaspore bauxite changes via a zone consisting of diaspore and boehmite into pure boehmite bauxite. It is interesting to observe that in the transitional zone the most common diaspore/boehmite ratio is 20/80.

The boundary between the transitional zone and the diaspore zone is sharp. Apparently the reaction boehmite → diaspore was in progress more rapidly at higher rates of nuclei formation.

Hematite is the predominant iron mineral throughout the deposits. The iron was fixed as goethite (probably secondary) only in the upper sections or in areas with poor drainage at the floor of the deposit.

Textural features of bauxites. Boehmite bauxites break up into fragments with smooth surfaces. Diaspore bauxites, which are much harder, break up irregularly into shelly pieces with rough surfaces. The bauxites are composed of concretions, pisolites and relics in a cement matrix. The pisolites are secondary textures formed

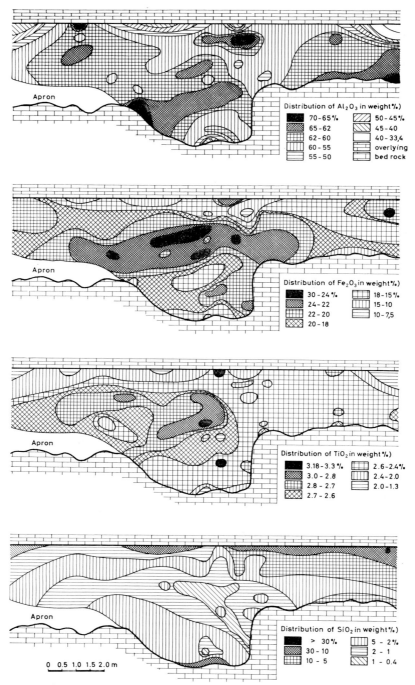

Fig.71. Distribution of major elements in the Late Cretaceous bauxites of the Kethro Mine (Parnassus, Greece). The highest alumina values and the lowest silica content are found in central parts of the dolina where drainage is at its maximum. (After Nia, 1968.)

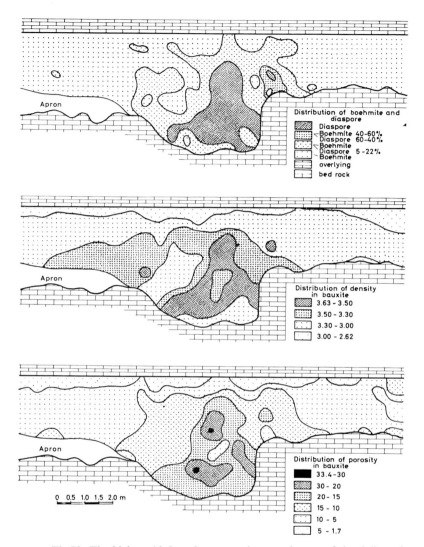

Fig.72. The highest Al₂O₃ values occur in central parts of the dolina where the highest concentration of diaspore is recorded. It was found that highest diaspore content coincides with highest density and greatest porosity. (After NIA, 1968.)

during very early stages of diagenesis. Along sineresis cracks the pisolites broke up and floated apart in the viscous matrix. This process of fragmentation is distinct from mechanical destruction which occurs during reworking and transport. There are two kinds of concretions:

(*1*) Uniformly opaque relic concretions, which are relics from early stages of diagenesis. Dissolution of iron progresses from localities of highest permeability.

There is a succession of light-coloured pisolites in a light-coloured matrix, dark-coloured pisolites in a light-coloured matrix and finally dark pisolites in a dark matrix. Angular to fringed relics typify this process (Plate VIII, 29).

(2) Concretions with light-coloured centres and opaque crusts in which faint pisolitic textures may be observed. They are typical for the transitional zone of boehmite and diaspore and the pure boehmite facies (Plate VIII, 30).

They are flattened, usually horizontally orientated in the upper zone, and low in iron. These secondary concretions grew from precipitation of iron on nuclei from solutions permeating the plastic material.

In the boehmite zone crystallites are submicroscopic, and iron and alumina minerals are intimately intergrown. The diaspore facies is coarser grained with a better separation of alumina from iron.

Diaspore occurs: (1) as coarse-grained growth zones in pisolites, often growing from the pisolite into the matrix; the pisolites are smaller than in the boehmite zone (Plate IX, 36); (2) as coarse crystalline secondary mineral formations in the matrix; (3) in opaque concretions; (4) as veinlets in the matrix, pisolites and concretions (Plate IX, 37).

In the transitional zone towards the boehmite facies diaspore fades gradually, first in the pisolites, then in the matrix and the concretions, though diaspore veinlets also reach far into the boehmite zone.

The following interpretation of genesis is given by NIA (1968): At first boehmite bauxite with boehmite pisolites developed. At a later stage, diaspore formed in the centre of the dolinas and at the same time boehmite formed in the marginal parts because of different drainage. Marginal boehmite pisolites grew bigger, but those of the centre changed into diaspore. During the transformation from boehmite into diaspore which was accompanied by separation of iron the pisolites became viscous and partly flattened. The transformation into diaspore resulted in a higher porosity in the central parts of the dolinas. The higher porosity and better drainage leading to fast removal of SiO_2 and separation of alumina from iron, rather than pH or Eh conditions, obviously favoured the formation of diaspore nuclei.

Hematite recrystallized during this process. The diagenetic process ceased because drainage stopped with a rising water table initiated by a marine transgression.

Missouri and Pennsylvania, U.S.A.

Palaeozoic diaspore bauxites like those of the U.S.A., the Urals and China may exhibit a regional facies differentiation.

The Cheltenham clay of Pennsylvanian age in Missouri was investigated in detail both in the field and in the laboratory by ALLEN (1935, 1952); MCQUEEN (1943); KELLER (1952); and KELLER et al. (1954), and in Pennsylvania by FOOSE

(1944); BOLGER and WEITZ (1951, 1952), and HUDDLE and PATTERSON (1961).

A widespread layer of the Graydon Formation (earliest Pennsylvanian) with cherty conglomerate breccia and quartzitic, pyritic sandstone rests unconformably on karst of the old land surface. Extreme irregularity in depositional features with lenses, slumps and soft-rock deformation typifies the formation. It is of the ancient stream channel deposit type. The Graydon Formation in turn is overlain by the Cheltenham fire clays, dipping gently in northeasterly, northerly and westerly directions from the Ozark dome (Fig.73).

The floor of the Cheltenham Formation is highly irregular and pockety. There are solution basins and sinkholes and occasional collapsed structures in the limestone beneath. In the southern plateaus karst penetrates 70 m, but the relief is much shallower (approx. 20 m) in the north.

Accordingly, the water table was at much lower levels in the south than in the north. The sediments are thickest in the irregular basins of the northern areas (about 20 m).

The Missouri fire clays grade from sandstones, plastic clay, semiplastic clay, semiflint clay and birly clay into diaspore. Red or green iron-bearing strata are also associated with them.

In the surroundings of the highland of the Ozark dome the Cheltenham Formation splits from south to the north into: (1) diaspore and flint clays, at the highest levels; (2) flint clay chiefly; (3) semiplastic and semiflint clays at the lowest levels (Fig.73A).

In KELLER's (1952) opinion changes in the lithofacies clearly represent distinguishable, lithologic types, which indicate the environment of formation.

At the generally even surface of the Cheltenham Formation, erosion and deposition reoccurred several times: (1) during the Pennsylvanian after lithification but prior to cover with younger strata; (2) prior to deposition of Tertiary(?) "hard-pan" gravels; (3) in Pleistocene and Recent.

The Cheltenham Formation occupied higher levels in the south than in the north. Therefore erosion occurred in lower horizons of this formation in the south (Fig.73B).

(1) Facies of semiplastic and flint clay. These occur as a blanket-type deposit the thickness of which ranges from about 20–1 m, averaging 7 m. Fire clay varies significantly from pit to pit much as coal lenses appear or fade. Apparently there were many local divisions into marshes, lagoons, hills and streams during sedimentary deposition. Mineralogically the fire clay in the lower part is composed of a mixture of more or less disordered kaolinite, some illite and quartz. The lattice order of fire clay crystals appears to increase, forming well-crystallized kaolinite as the clay becomes "harder", and approaches flint clays in the upper part. Intensive weathering leached soluble cations.

The texture of the semiplastic and semiflint clay is commonly that of

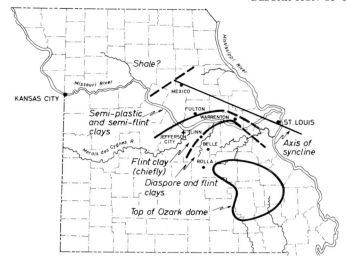

A

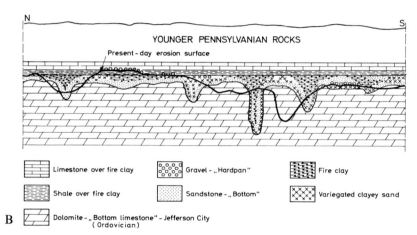

B

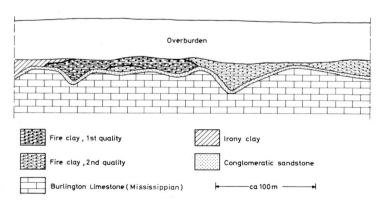

C

Fig. 73.

random (criss-cross) orientation interpreted as the growth of crystal aggregates of clay minerals from homogeneous mud or gel. The isotropic texture is seen as the result of flocculation from water containing a high concentration of effective cations. The crystal growth may be explained by aggradation or by degradation (weathering of the primary silicates).

It is inferred that this facies developed in a low-lying marshy terrain close to the sea. The land surface on the limestone was irregular with low to moderate relief, with karst depressions undrained by subsurface streams, attesting to slump features, and structures shown in clayey coal and light to dark clay strata. The soil and the bedrock would have been waterlogged. Slightly acid pore water moving upwards (O_2, organic acid) leached Na^+, Ca^{2+}, Mg^{2+} and K^+ and removed silica to some extent (Table XXVIII). Seaward from this belt a typical marine shale facies probably developed. There is a thickening to the north in the direction of the facies change.

(2) The flint clay belt. In the southerly direction the semiplastic and semiflint clays gradually change into flint clay which occupies slightly higher levels on the flanks of the Ozark dome. The only remnants of pre-Pleistocene denudation are deep karst depressions in the form of isolated sinkholes.

An increase in the lattice order of kaolinite and a decrease in illite content are the main mineralogical differences between flint and semiplastic clays. Anatase

---

Fig.73. A. Map of Missouri showing belts within the Pennsylvanian rocks which are characterized by particular clay types, i.e. litho-facies types within the Cheltenham. The alumina content of the clays typifying each band increases progressively toward the apex of the Ozark dome. The belts are roughly concentric in the northwest sector of that flank of the dome, and they terminate where erosion has removed the Pennsylvanian. The axis of the syncline in which flint clay, semiflint, and semiplastic fire clays were preserved is shown between St. Louis and Moberly. The more plastic fire clays occur lowest in the syncline; flint clay occurs in the northeast flank (Whiteside locality) where the rocks rise on the Lincoln fold, and toward the south where they were raised by the Ozark doming.

B. A diagrammatic cross-section through a part of the diaspore-flint clay district of Missouri. The complete section shows the stratigraphy as it occurred at the Bueker Pit shortly after deposition of the Fort Scott limestone (over the fire clay). The heavy line which cuts irregularly across the dolomite and the overlying rocks (Pennsylvanian) represents the surface of gravel, shown in two places, indicating prior erosion followed by deposition of a sheet of gravel, and subsequently renewed erosion.

Below the present erosion surface are found Pennsylvanian sediments devoid of fire clay; shallow deposits of fire clay, and a few deep deposits filled with fire clay and/or sandy, "worthless" clay. The deep deposits are preserved as buried remnants of erosion. They may be surrounded by Pennsylvanian rocks or by Ordovician dolomite, depending upon the vagaries of erosion as occurred in Cheltenham and Recent times.

C. A diagrammatic cross-section through a semiplastic and semiflint fire clay pit modified slightly (simplified) from the actual occurrence. Note that the distribution of high-quality fire clay in the deposit is independent of stratigraphic control; it is apparently random within one pit and between two or more pits. Source material and local environment were variable during any one time. (After KELLER et al., 1954.)

TABLE XXVIII

CHEMICAL ANALYSES OF MISSOURI FIRE CLAYS (after McQUEEN, 1943)

	1	2	3	4	5
SiO_2	3.89	44.42	45.92	48.90	56.10
Al_2O_3	76.21	38.63	35.79	33.20	24.47
Fe_2O_3	0.98	0.55	0.75	1.47	3.64
TiO_2	3.52	2.12	2.28	1.58	1.58
CaO	0.08	0.04	0.06	0.56	0.61
MgO	0.06	0.10	0.36	0.34	1.11
Na_2O	0.79	0.30	0.44	0.10	0.17
K_2O	0.24	0.12	0.41	1.53	2.89
H_2O	14.56	13.90	13.06	11.55	8.39
Total	100.33	100.18	99.07	99.29	99.27

1. Diaspore clay, $NE^1/_4$ $SE^1/_4$, Sec. 29, T. 41 N., R. 7 W., Maries County, Mo.; *2.* Flint fire clay, $SE^1/_4$, Sec. 19, T. 51 N., R. 1 W., Lincoln County, Mo.; *3.* Semi-flint fire clay, $NW^1/_4$, Sec. 29, T. 47 N., R. 3 W., Warren County, Mo.; *4.* Semi-plastic fire clay, A. P. Green Fire Brick Co., Pit No. 4, Audrain County, Mo.; *5.* Plastic "foundry" clay, A. P. Green Fire Brick Co., Pit No. 4, Audrain County, Mo.

accounts for TiO_2. Flint clay may range in colour from black to white. From coal and sparse pyrites, it may be concluded that negative Eh values prevailed in the basin. The flint clay belt was a more stable land area with a higher H^+ and lower K^+ concentration.

(*3*) The diaspore clay belt. The diaspore (and boehmite) and the intermediate diaspore kaolinite clays occur in sinkholes of different size closest to the summit of the Ozark dome. Lenticular layers of diaspore occur in the lower two thirds of the deposits and dip gently towards the centre of the sinkholes. The surrounding flint clay is of prime quality.

Highest of all in alumina are the diaspore clays (Table XXVIII). Therefore, leaching of silica and/or the concentration of alumina were at a maximum. A close correlation between occurrence and composition of the diaspore clay is inferred.

Diaspore clays have a high permeability compared with flint clays, and the boehmite clays measured generally have a permeability intermediate to that of diaspore and that of flint clays. The apparent increase in permeability with increase in alumina content suggests that permeability has increased as silica was removed from flint clay to form boehmite and diaspore clays (ALLEN, 1954).

Mineralogically the diaspore clay is composed of diaspore and well-crystallized kaolinite. Boehmite occurs in place of diaspore in a few deposits. Goethite, lepidocrocite and anatase are also commonly present. There is a complete range from pure flint clay to pure diaspore. Gibbsite has not been found as primary mineral but as a product of presentday weathering.

From the association of diaspore with coal, organic carbon and occasionally with pyrite, KELLER (1952) derives evidence that diaspore formed under reducing conditions in a water-logged environment. Granular textures (oolites, pisolites) may have developed directly from gels.

Diaspore grains are definitely not detrital; this is demonstrated by: (*1*) interfingering of diaspore and flint at any angle to the horizontal; (*2*) continuously graded boundaries between them; (*3*) microscopic textures which are unlike the boundaries of detrital particles.

ALLEN (1935) describes the progressive change of oolites composed of kaolinite to those consisting entirely of diaspore. Sometimes boehmite and diaspore are associated in this facies. Diaspore, characteristically is more porous than boehmite, and it is doubtful whether an increase in density from about 3.0 of boehmite to about 3.4 for diaspore could account for the increased porosity.

Boehmite forms as a fine-grained matrix and was never observed to develop oolitic textures like diaspore. There is no evidence in the field to assume that diaspore originates from pre-existing boehmite. The genetic relationship of boehmite and diaspore is not understood. The flint and diaspore clay developed under essentially the same conditions.

BOLGER and WEITZ (1952) show that the fire clay (Mercer Formation) of the Potsville series of Pennsylvanian age formed under the same conditions as those demonstrated by KELLER (1952) for the Cheltenham clay. The main minerals of the Mercer Formation are kaolinite and diaspore with subordinate boehmite, iron oxides, iron sulphides, siderite, and mica-like clay minerals. Microscopic investigations showed repeated crystallization of the various minerals in matrix, concretions, nodules and veinlets.

A common transport of fine-grained matrix, pisolites and nodules may be excluded. Fragments of drifted diaspore nodules also indicate diaspore formation during the early stages of diagenesis. The characteristic environments were swamps, too, supplied with colloidal gels from sluggish streams. Kaolinite, boehmite and diaspore crystallized from the gels. After lithification, crystallization in cracks caused by shrinkage continued over a long period. Kaolinite and diaspore formed in the cracks, replacing each other in succession. Hardening occurred prior to post-fire-clay sedimentation.

The Urals

According to GLADKOVSKY and USHATINSKY (1963) Eifelian bauxite in the northern Ural Basin demonstrates eluvial ore formation. The relationship of boehmite bauxite to diaspore bauxite can not be explained by either sedimentation or metamorphism. In the sinkholes both the red bauxite in the lowest part and the grey bauxite in the topmost part consist chiefly of diaspore and boehmite. The central part of red-soiling and non-soiling bauxite consists of diagenetic diaspore in the pisolites and boehmite in the cement. The boehmite in the surrounding bauxite has not changed into diaspore.

China

In the Gun district in the Honan Province of China (SCHÜLLER, 1957) karstificated Ordovician limestones are overlain by one oolitic chamosite horizon with idioblastic siderite or greyish-green chloritic slates with diaspore oolites; laterally and vertically this facies passes into kaolinite bauxites or hard grey bauxites. In the top section the bauxites become brown or brownish black from carbonaceous matter and are overlain by coal. The profile described is succeeded by fusulinid limestones of the lower section of Late Carboniferous. The source rock of these bauxites is a chloritic to muscovitic slate. The relic heavy minerals are tourmaline and zircon. Quartz is absent. The bauxites do not contain iron oxides. Also boehmite is absent in contrast to the bauxites of Greece, the U.S.A. and Russia. Relic pisolites and lumps indicate repeated reworking and redeposition of the still plastic sediment. Angular shapes of kaolinite encrusted with muscovite indicate feldspar relics. Besides detrital layer silicates in the matrix, new minerals formed such as chlorite, muscovite, vermicular kaolinite and diaspore $(5–10\ \mu)$. Diaspore also occurs in more or less shelly oolites as idiomorphous crystals or alveolar intergrown and in a coarse crystalline $(20–40\ \mu)$ form. These bauxites contain 10–20% SiO_2, 50–70% Al_2O_3, 2–4% TiO_2, very little iron, organic carbon up to 3%, $> 1\%$ K_2O and $< 1\%$ Na_2O.

The abnormally high content of the trace elements Cr, V and Ni indicate basic source rocks.

Our interpretation of diagenesis is:

(1) In most cases the bauxites rest solely on karst bedrock but did not develop from carbonate rocks.

(2) The bauxites formed from repeatedly reworked material.

(3) Sedimentation commenced with chamosite as the dominant mineral (also iron, chlorite and siderite) and ceased with minerals rich in Al. Finally, peat bog developed.

(4) The lateral facies change is from bauxites to kaolinitic bauxites to shales to anthracite.

(5) Diaspore and kaolinite are contemporaneous syngenetic mineral phases and formed both in vertical and horizontal directions. There was no clay formation at a later stage.

(6) During diagenesis diaspore pisolites and diaspore concretions developed. Chlorite, muscovite and diaspore are not of metamorphic origin.

Kashmir, India

The following profile developed in Jammu Kashmir (RAN, 1931):

Eocene: coal
                   diaspore bauxites, 2–3 m thick
                   siliceous breccia
                   limestones (inferred)
pre-Triassic: limestones

Laterally the diasporites change into clay seams. The sediments are solid, massive or pisolitic, frequently with fine stratification, off-white to grey in colour, contaminated by organic matter.

The pisolites were deformed and elongated in the plastic stage of the sediment. The matrix consists mainly of submicroscopic boehmite. Diaspore occurs as bigger crystals in matrix and pisolites. The sediments are characterized by low Si and Fe values and a high alumina content. They are contaminated by organic carbon.

Chemical analyses of typical specimens from the Jammu deposit are given in Table XXIX.

TABLE XXIX

CHEMICAL ANALYSES OF TYPICAL SPECIMENS FROM THE JAMMU DEPOSIT

	Pisolite (west of Jangal)	Chakar (C.d; 3rd ft.)	Chakar C.s; 1st ft.)	Chakar (C.v; 1st ft.)	Chakar (D.c; 2nd ft.)	Chakar (D.e; 3rd ft.)
SiO_2	0.84	6.78	6.53	14.07	0.79	1.31
Fe_2O_3	1.09	3.56	1.68	3.88	0.75	1.86
Al_2O_3	78.34	70.71	73.43	61.70	80.74	78.24
MgO	0.13	0.05	0.02	0.03	0.77	0.04
CaO	0.20	0.02	0.10	0.22	0.10	0.21
K_2O	nil	nil	0.24	nil	n.d.	0.15
Na_2O	nil	nil	0.42	0.56	n.d.	nil
H_2O^+	14.99	14.37	13.87	14.48	12.15	14.59
H_2O^-	0.03	0.71	0.21	0.39	0.11	0.13
TiO_2	4.35	4.36	3.08	3.90	3.38	2.48
C	0.22	0.12	0.11	0.82	1.46	1.35
	100.19	100.68	99.69	100.05	100.25	100.36

All analyses of diaspore bauxites obtained so far are plotted in Fig.74 and 75.

Three groups of diaspore bauxites may be distinguished, i.e.:

group 1: along the line SiO_2–Al_2O_3 (Missouri, Honan, Jammu)

group 2: along the line Fe_2O_3–Al_2O_3 (Greece)

group 3: in the inner part of the field with additional minerals of bivalent iron (France)

In all cases diaspore seems to develop at the expense of boehmite during late

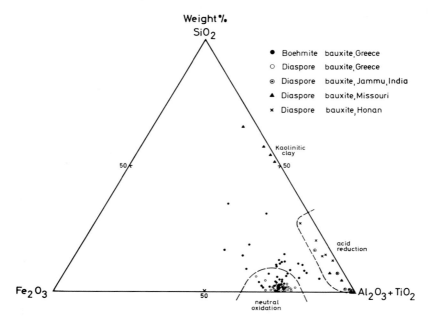

Fig.74. Triangular projection (SiO_2–Fe_2O_3–Al_2O_3) of chemical analyses of boehmite and diaspore bauxites. The diaspore bauxites or diaspore clays of Missouri, Honan and Jammu are projected close to the SiO_2–Al_2O_3 line. The diaspore bauxites of Greece are located close to the Al_2O_3–Fe_2O_3 line, while Greek boehmite bauxites also lie in the inner part of the field.

stages of diagenesis. Boehmite bauxites show intimate and submicroscopic inter-growth of boehmite and hematite. During transformation from boehmite to diaspore observations of ore show:

group 1: iron dissolved and migrated from the horizon;

group 2: iron dissolved and precipitated to form well-crystallized hematite;

group 3: iron dissolved and minerals of bivalent iron such as siderite or pyrite formed.

In all cases spatial separation of iron and alumina took place. Both the Al phase and the Fe phase became coarse crystalline. The process of recrystallization was caused by a high rate of dissolution possible only in areas with optimum drainage. pH and Eh conditions probably varied. In the case of Greek bauxites Fe^{3+} minerals reprecipitated, indicating a neutral and oxidizing environment. In the case of bauxites of Missouri, Honan and Jammu Fe^{2+} migrated from the ore body. In the bauxites of France siderite associated with diaspore precipitated from Fe^{2+} solution, indicating an acidic and reducing environment in both cases.

Summary. Diagenesis of bauxite formation progresses in several stages in a terrestrial alkaline or neutral environment.

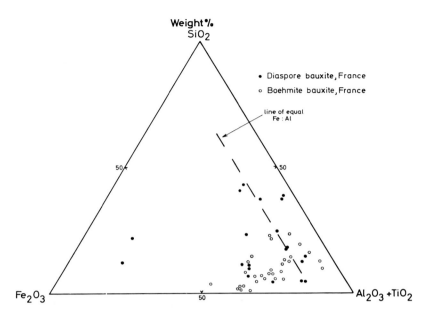

Fig.75. Diaspore bauxites of France, like boehmite bauxites, are projected into the inner part of the field. In all cases iron was mobilized and migrated from the bauxite or reprecipitated as hematite or siderite.

(*1*) All processes are a function of: (*a*) the drainage pattern governed by the joint system of the karst bedrock; (*b*) the vegetation.

(*2*) The initial step is desilicification which is strongest in areas with optimum drainage. Three-layer minerals transform into two-layer minerals (kaolinite group) which in turn change into hydroxides of alumina. Primary Al minerals are gibbsite or boehmite and primary Fe minerals are mostly hematite.

(*3*) Neomineralization of kaolinite is caused by silicification of the top section of the profile; iron dissolves but frequently reprecipitates as goethite.

(*4*) In areas with optimum drainage gibbsite may be dehydrated to form boehmite. In fine-grained boehmite-hematite bauxites, there is a contemporaneous crystallisation of boehmite and hematite from gel. Under conditions of strong drainage boehmite may change to diaspore. The resulting separation of iron and alumina leads to recrystallization of hematite in the matrix or removal of iron. Some diaspore bauxites are nearly devoid of iron minerals. In some Palaeozoic and Mesozoic occurrences formation of diaspore seems to follow the separation of iron and alumina. Dehydration to corundum is described from bauxites in Russia and Hungary.

(*5*) The following textures may result from neomineralization: (*a*) concentric pisolites, 0.2–20 mm in diameter; (*b*) relic concretions resulting from the dissolution of iron minerals in the surrounding matrix; (*c*) impregnation concretions resulting from iron incrustation (hematite or goethite) of bauxite sections originally rich in iron; (*d*) consistency rock is earthy or plastic; for this reason flow textures frequently develop which may contain deformed pisolites with preferred orientation.

(*6*) Pisolites and impregnation concretions are the first to dehydrate to diaspore but mineral transformation of the matrix mainly occurs at a later stage. During diagenesis certain trace elements such as Mn, Ni, Co dissolve and concentrate through precipitation at the base of the bauxite deposit.

Post-diagenetic environment

Post-diagenetic processes may be grouped as follows: (*1*) epigenetic alterations caused by reaction with pore waters from overlying strata; (*2*) post-uplift weathering and erosion of capping sediments at the present-day surface.

Both processes may lead to convergent paragenesis. The most important epigenetic alterations are rehydration, resilicification, deferrification, sideritization, pyrification, calcitification.

Following bauxite genesis many areas, particularly unstable shelf regions, develop terrestrial or coastal swamps with an overproduction of plants. A reducing and acidic environment in the underlying bauxites may result from a water-logged condition and an influx of humic acid. Examples are bauxite deposits in the U.S.S.R., Yugoslavia, Greece, France and Hungary. The Hungarian bauxites (VADAZ, 1951; KOMLOSSY, 1968), overlain by lignites and lignitic clays, developed from extensive swamps during Eocene times and suffered strong alterations.

Considerable epigenetic resilicification and rehydration rook place both in Carboniferous deposits (Tikhvin) and in the Mesozoic to Cenozoic bauxites in the U.S.S.R. (Transurals, Kasakstan; GLADKOVSKY and USHATINSKY, 1963). The iron minerals in particular but also alumina minerals decomposed. From the overlying lignitic seams bauxite was supplied with S, P, CO_2 and SiO_2 and disintegrated, particularly at the surface but also along joints in deeper sections. The initial process was the mobilization of iron. Depending on ground water movements iron minerals dissolved, and light coloured, so-called bleached bauxites with low iron content were left behind. Otherwise thick massive pyrite or markasite lenses at the top or bottom boundaries or streaks in the bauxite body itself formed from reaction with S^{2-}.

In the grey bauxites pyrite often occurs disseminated in the matrix or is found to be pseudomorph in pisolites and concretions.

In the presence of CO_2 siderite formed. It may replace the iron-rich zones in pisolites or concretions. Occasionally it also formed big concretions in the matrix or filled pore space, joints and fissures. Similarly Ca-Mg-carbonates may fill the pore space or initiate replacements.

The epigenetic processes are characterized by rehydration, mobilization and the removal of Si and Fe or neomineralization of kaolinite, iron carbonates or iron sulphides.

Weathering following uplift progresses mainly in an *acid and oxidizing environment*, as seen especially in Hungarian bauxites. Weathering is characterized by the dissolution and neomineralization of sulphates, aluminates and phosphates. Re-oxidation of the upper bleached bauxites frequently taking place in the upward direction disintegrates sulphides (KOMLOSSY, 1968). Pyrite concretions are encrusted with hematite or goethite, which may dissolve again at a later stage. Melanterite, $Fe[SO_4] \cdot 7 H_2O$, also results from the disintegration of pyrite (BARDOSSY, 1954). Decomposition of sulphides in the upper section of the bauxite body may lead to secondary impregnation of the lower part of the deposit with sulphide solutions, i.e., descending pyritization (KOMLOSSY, 1968).

During reoxidation, the Ca and SO_4 precipitate to form gypsum $CaSO_4 \cdot 2 H_2O$, and Al reacts to form alunite $K Al_3 [(OH)_6 (SO_4)_2]$ (VADAZ, 1943) and augelite $Al_2[(OH)_3/PO_4]$. These neomineralizations fill pore spaces or form coatings on joints and in cavities.

Terra rossa formation along the top and bottom boundaries of bauxite deposits occurring near the surface may be so intense as to cause sinking of bauxite and capping sediments. The bauxite breaks up, resulting in dissolution and neomineralization of kaolinite and goethite along joints. During very strong post-genetic karstification bauxite degrades to earthy red and yellow or white clayey bauxite (Plate VII, 26). Such examples are found in many places in France, Hungary and Greece.

Summary

(*1*) In a reducing acidic environment Fe^{3+} minerals dissolve, and iron may be reprecipitated as pyrite or siderite. Other possible reactions are rehydration of Al and Fe minerals and kaolinitization.

(*2*) In an oxidizing environment sulphates, aluminates and phosphates may form.

(*3*) The bauxite may regain an earthy soft consistency by post-genetic karstification of bedrock and terra rossa formation on bauxite. The only upgrading factor is removal of iron; all other post-diagenetic alterations lower the quality of the primary bauxites.

"Bauxites" on phosphate rocks

From Taiba in Senegal, CAP DE COMME (1952), SLANSKY et al. (1964) and TRESSLER (1965) describe phosphorous lateritoides which rest on a clay succession of Early Lutetian age. The profile recorded is:

(4) Al-phosphates: alternating grey, off-white, white and greenish lithified

layers, 3–4 m thick, porous with many cavities; frequently a reworked upper section mixed with capping sands is noted.

(3) Clay seams: brown or multicoloured, ranging from 0.5–2 m due to sub-aqueous sliding.

(2) Ca-phosphates: conglomeratic phosphate layers variable in thickness alternating with sand, silt and chert; reworked marine horizon clays: off-white clays with fine stratification.

(1) Underbed.

Following uplift fluvial erosion left behind irregular surfaces.

Mineralogically and chemically the four horizons are composed of: (*1*) The underlying clays consist of varying proportions of montmorillonite and attapulgite. (*2*) In the layers rich in P the Ca-phosphate succession is composed of oolitic or pseudo-oolitic fluorapatite. The clays in the phosphate layers and the clay seams consist of montmorillonite with minor amounts of kaolinite. (*3*) The interbedded clay seams are made of montmorillonite. There may be traces of kaolinite. (*4*) Topmost part contains Al-phosphates (Table XXX).

While the sediments of the lower profile are unaltered, the top section has undergone lateritization with removal of Ca, Fe and Si.

Ca-phosphates, relic textures of which are still preserved, transformed into Al-phosphates. In the zone of Al-phosphates crandallite $CaAl_3H[(OH)_6/(PO_4)_2]$ is dominant with minor amounts of augelite $Al_2[(OH_3/PO_4]$ occurring particularly in the upper section. Instead of montmorillonite kaolinite and illite formed. The chert layers pulverized or dissolved.

There is an ambiguity as to whether enrichments of iron in the clays underlying the phosphates are primary or result from lateritization. The frontal line of lateritization is therefore undefined.

TABLE XXX

CHEMICAL COMPOSITION OF THE ZONE OF AL-PHOSPHATES (after SLANSKY et al., 1964)

Sample	P_2O_5	CaO	Al_2O_3	Fe_2O_3	CO_2	SiO_2	MnO	CaO/P_2O_5
78	28.7	6.4	31.6	2.4	0.2	10.8	0.5	0.22
77	31.7	7.0	32.3	2.8	0.4	3.8	0.35	0.72
76	28.7	9.8	31.5	2.0	0.5	6.8	0.80	0.34
75	25.7	8.8	28.6	2.4	0.4	15.9	0.85	0.34
74	28.3	8	31.2	2.8	0.7	10.1	1.0	0.28
73	15.1	5.6	21.1	3.6	0.2	41.3	0.9	0.36
72	19.9	6.6	25.2	2.8	0.4	29.1	1.05	0.33
71	13.5	5.5	22.8	6.0	0.5	38.3	1.05	0.40
70	13.4	5.6	18.7	11.6	0.4	38.4	1.45	0.41
69	15.0	3.2	14.2	5.2	0.5	52.4	0.6	0.21
68	20.8	3.6	21.6	18.0	0.2	17.1	0.9	0.17

Chapter 6

GEOCHEMISTRY OF BAUXITE DEPOSITS

This chapter is concerned with the general geochemical laws of bauxite-deposit formation. Reference to Chapter 5 will show that the chemical composition of bauxite deposits is governed by:

(*1*) the amounts of specific elements in the source rock; (*2*) the chemical association of specific elements with stable or unstable minerals during weathering; (*3*) the intensity of drainage during weathering: (*a*) precipitation in situ (relative enrichment); (*b*) vertical or horizontal ground water transport of elements (absolute enrichment); (*4*) polygenetic alterations: (*a*) late diagenetic; (*b*) epigenetic.

The important elements in bauxite genesis are Si, Al and Fe. Extensive deposits exclusively or partly formed by the removal of silica, resulting in relative enrichment of Al and Fe. These bauxites are characterized by the Al/Fe ratios corresponding to those of the source rocks. Such ferrallites are certain horizons of bauxites on igneous rocks and most of the karst deposits. Siallites (rocks rich in silica) develop if iron is removed faster than silica. They form the saprolite zone on igneous rocks and develop highly aluminous clays and transitional stages to flint clay on sediments. Allite formation occurs by relative enrichment of aluminium through the selective removal of silica and iron.

Besides relative enrichment, absolute enrichment—as impregnations—may occur, resulting from colloidal or ionic transport of Fe, Al and Si in ground water over long distances vertically and horizontally, followed by reprecipitation. (For examples of such deposits refer to Chapter 5.) Basic laws are considered to underly the phenomena.

The relative enrichment of trace elements in bauxite with respect to the earth's crust is shown in Table XXXI (after BENESLAVSKY, 1963).

SILICA

During early stages of diagenesis enrichment of silica is the reciprocal of that of aluminium. There is a direct relationship between the silica removal and intensity of drainage in the bauxite. For this reason many bauxites show gradual lateral and vertical facies changes from siallites (40% SiO_2) into allites or ferrallites (2% SiO_2).

In plateau bauxites on igneous rocks, silica is highest in the lowest and

TABLE XXXI

CONCENTRATION OF ELEMENTS IN THE EARTH'S CRUST AND IN BAUXITES

Elements	Bauxites	Earth's crust	Concentration
Al	30.16	7.45	4.05
Fe	9.99	4.20	2.38
Si	3.80	26.00	0.15
Ti	1.64	0.64	2.56
Ca	0.81	3.25	0.25
C	0.515	0.35	—
S	0.50	0.10	5.00
K	0.36	2.35	0.15
P	0.13	0.12	1.08
Mg	0.09	2.35	0.038
Na	0.08	2.40	0.03
Mn	0.076	0.10	0.76
V	0.062	0.02	3.10
Cr	0.055	0.03	1.83
Cu	0.023	0.01	2.30
Zn	0.021	0.02	1.05
Ga	0.0035	—	—

uppermost zones. There is resilicification in situ of the basal saprolite zone caused by large amounts of silica freed from decomposing tectosilicates above, while water-consuming vegetation may be responsible for SiO_2 fixation in the topmost zone, leading to clay-mineral formation. As all fully preserved bauxite deposits show a similar distribution of SiO_2, general laws are thought to govern the SiO_2 spread.

Furthermore, a direct relationship between SiO_2 distribution and intensity of drainage is also observed in bauxites altered by diagenesis and interbedded with sediments (Jamaica, Greece, Surinam). The uppermost part is also characterized by strong resilicification, probably resulting from dehydration through growth of vegetation. The neomineralization consists of the kaolinite group, but when the ground is water-logged minerals of the montmorillonite group form. There is no quartz crystallization in bauxite deposits. Impregnation of fissures and pore space or replacement of bauxite by clay minerals resulting from epigenetic SiO_2 mobilization may both occur.

ALUMINIUM

An increasing degree of drainage, increasing speed of SiO_2 removal, and increasing relative enrichment of aluminium govern the early stages of diagenesis. Highly porous rocks with good preservation of relic textures are mainly areas of relative

Al enrichment in the plateau bauxites, with gibbsite being the major Al-mineral. Drainage intensity increases vertically towards the top and laterally towards the valley edges. These areas of optimum drainage are characterized by absolute Al-enrichment, pisolite formation and the occurrence of boehmite in addition to gibbsite. As yet there are no exact data on the extent of vertical and lateral Al migration.

The aluminium may be enriched by 300–400 % with respect to the source rock. The highest Al_2O_3 values recorded are from the central parts of deposits (profiles from Arkansas, India, Australia) and range from 60–70 % Al_2O_3.

Epigenetic aluminium migration is subordinate and only occasionally recrystallization or gibbsite filling of pore space to a greater extent is observed in plateau bauxites.

In bauxites interbedded with sediments, the highest Al_2O_3 values recorded also coincide with the drainage optimum. Gibbsite or boehmite are reported to be primary minerals. There is increasing pisolite formation with increasing drainage intensity in the plateau bauxites, yielding the highest Al content. Gibbsite dehydrates to boehmite or diaspore. The highest aluminium content is confined to the central parts of the deposits with 80 % Al_2O_3 in diaspore bauxites.

Boehmite bauxites are always characterized by submicroscopic grain sizes, while gibbsite bauxites may be fine crystalline or coarse grained (up to 100 μ); diaspore bauxites are always coarse crystalline. Commonly there is submicroscopic intimate intergrowth of Fe and Al minerals in gibbsite and boehmite bauxites, but the separation of Fe and Al and porosity improve with increasing intensity of drainage. All these phenomena indicate a mainly relative Al-enrichment during diagenesis of karst bauxites, while Al-migration and absolute Al-enrichment are subordinate.

Several trace elements, such as Ti, Zr and V, follow the pattern of Al enrichment (see section "Trace elements", p.194).

IRON

In bauxites on igneous rocks there may be both relative and absolute enrichments of iron. Relative enrichments in ferrallites reflect the Al/Fe ratios of source rocks. There is absolute enrichment by iron impregnation of the upper iron crust and in many parts of the saprolite. While iron values may be as low as $< 1 \% Fe_2O_3$ in central parts of the profiles, they may reach a maximum of 50–60 % Fe_2O_3 in the iron cap. The common primary iron mineral is hematite, formed from gel. In horizons with absolute iron enrichment and crystallization from gels, iron and siallite components separate to form vesicular textures with hematite on the one hand and kaolinite or gibbsite on the other. Goethite may also be primary (on charnockites in southern India) if iron enrichment is exclusively relative. Epigenetic iron minerals are goethite, rarely maghemite and siderite when water-logged conditions prevail.

In gibbsite and boehmite karst bauxites there is a very close intergrowth of Fe and Al minerals, forming ferrallites or fersiallites. Most bauxites on sediments and on basic igneous rocks have similar Al/Fe ratios. For this reason ferrallitic weathering products of basic igneous rocks may be regarded as source rocks of many bauxites on karst. The Fe_2O_3 content of red karst bauxites ranges from 20–30% Fe_2O_3, which means 300–400% enrichment with respect to source rocks. Al and Fe become separated during diagenesis, epigenesis and weathering. Secondary infiltration and iron enrichment only takes place locally. During diagenesis and epigenesis iron is commonly removed from the bulk of the bauxites. The Fe_2O_3 content does not exceed 1–2% in white bauxites.

The primary iron mineral is hematite, while secondary minerals are goethite and siderite, the latter being formed under waterlogged conditions.

TITANIUM

According to MIGDISOW (1960) titanium enrichment during weathering is highest in bauxites. The geochemical distribution pattern in such cases is:

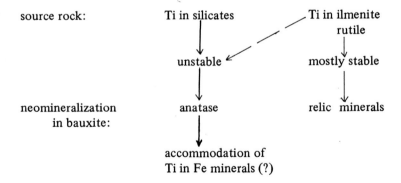

In bauxites on igneous and metamorphic rocks titanium in relic minerals is confined to rutile, titanomagnetite, ilmenite or in diagenetic phases, commonly to anatase. Like Al and Fe, Ti may become enriched relatively or absolutely by migration and precipitation from solution. The highest TiO_2 values recorded in bauxites range from 15 to 32%. Thus the titanium content is governed by the amount of Ti available in the source rocks and by the degree of mobilization (Table XXXII).

There is a largely parallel trend of increasing Al and Ti values, and in central

TABLE XXXII

COMPARISON OF TI CONTENT IN BAUXITES AND SOURCE ROCKS

Occurrence	Basalt, Gujerat (India)	Bauxite, Gujerat (India)	Basalt, Udagiri (India)	Bauxite, Udagiri (India)	Nepheline syenite, Saline Cy. (U.S.A.)	Zone of concretions, Saline Cy. (U.S.A.)	Charnockite, Kotagiri (India)	Bauxite 2, Kotagiri (India)	Charnockite, Yercaud II (India)	Bauxite 7, Yercaud II (India)	Andesite, Belian (Serawak)	Hill bauxite (average), Belian (Serawak)	Dacite, Manus Island (Australia)	Bauxite, Manus Island (Australia)
SiO_2	54.26	0.6	50.51	1.2	56.51	1–7	67.03	2.67	64.14	0.19	52.49	1.69	67.31	0.36
Al_2O_3	17.26	64.2	17.18	54.2	21.11	57–61	14.99	53.07	18.85	57.10	17.70	54.73	14.22	56.71
$Fe_2O_3 + FeO$	11.65	1.2	14.59	4.2	3.0	1–3.5	6.4	9.94	5.13	9.03	8.00	9.90	5.58	10.84
TiO_2	0.81	2.9	2.58	4.3	0.5	1.2–4.4	0.91	1.57	0.70	1.88	1.82	2.47	0.53	1.09
TiO_2 ratio source-rock/bauxite		3.6		1.7		2.4–8.7		1.7		2.7		1.4		2.1

parts of the profile the optimum Ti enrichment is 200–400% with respect to the source rocks. Titanium crystallized to submicroscopic dispersed anatase (Arkansas, Gujerat, etc.).

There may be absolute enrichment of titanium as titaniferous laminae and concretions if titanium precipitates from vertically or laterally moving ground water solutions. Aluminium and titanium have a very similar geochemical distribution pattern resulting in close relationships between Ti and Al values in many bauxite deposits.

In karst bauxites the mean Ti values range from 2–4% TiO_2. Ti is confined to submicroscopic anatase neomineralization to a great extent. By preference Ti follows Al enrichment, but in places there is an affinity to iron accumulation.

Corresponding Al/Ti ratios of clastic bauxites and source rocks exist, but a more pronounced correlation between Al and Ti is apparent in karst bauxites with repeated dissolution and precipitation during several stages of diagenesis. BARDOSSY and BARDOSSY (1954) described preferred enrichment of Ti with Al (Fig.76) but also referred to Ti-enrichment with Fe. There is selective enrichment of Ti with Al in the ferrallitic main horizon and a parallel enrichment with iron

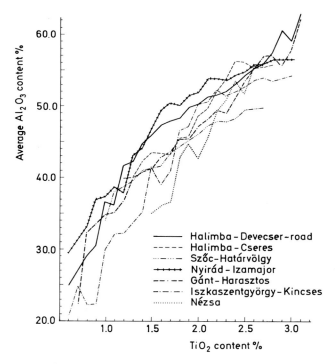

Fig.76. The variation of TiO_2 values as a function of Al_2O_3 content in Hungarian karst bauxites. (After BARDOSSY and BARDOSSY, 1954.)

in the siallitic top zone during diagenesis in bauxites of France (VALETON, 1965). In the ferrallite zone Ti precipitated as anatase, but it is associated with secondary iron concretions in the siallite zone.

TRACE ELEMENTS

The process of successive Si and Fe dislocation or removal leads to a continuous Al-enrichment in laterite bauxites and in bauxites on sediments. The phasic mobilization during diagenesis, epigenesis and younger weathering causes the trace elements to be displaced several times.

Certain trace elements may become enriched to such an extent as to render deposits mineable or are mined as by-products, like Ti, Ni, Co and Cu.

In principle trace elements show a similar geochemical behaviour in bauxites on igneous, metamorphic and sedimentary rocks.

The geochemical distribution pattern of various trace elements is demonstrated in bauxites on igneous and metamorphic rocks, and in karst bauxites.

Trace elements in bauxites on igneous and metamorphic rocks

GORDON et al. (1958) compared the trace element content of nepheline syenite in Arkansas, U.S.A., and four bauxite types. They observed Cr, Cu, Ga, Nb and Mo to be generally more strongly enriched than Al, while there is a less pronounced trend of Zr, Ti, Sc, V, Be, Mn, Y and Pb enrichment compared with Al.

Sr, La, Ba, Ca, Mg and the alkalis migrated from the bauxite (Tables XXXIII, XXXIV).

ADAMS and RICHARDSON (1960) investigated the distribution of several trace elements in a vertical profile of Arkansas bauxites and found Th, U and Zr to be enriched proportionally to Al (Fig.77).

A strong enrichment of gallium with aluminium in Indian bauxites on basalts was shown by CHOWDHURY et al. (1965).

WOLFENDEN (1965) showed for Malaysia that there is selective enrichment of certain elements as a function of environment during late diagenesis. The trace element content of andesites is given in Table XXXV. The elements enriched or removed differ in hill bauxites (Table XXXVI) with good drainage conditions and in swamp bauxites (Table XXXVII) or in the basal saprolite (Table XXXVIII) where waterlogged conditions prevail.

In the saprolite the elements Cr and Zr become enriched relatively due to their association with chromite and zircon which are resistant to weathering, while Ni, Co and P concentrate via solution (refer to Ni- or phosphate bauxites, p.195).

Ga in particular precipitates in a reducing environment and becomes en-

TABLE XXXIII

AVERAGES OF SPECTROGRAPHIC DETERMINATION OF TRACE ELEMENTS IN NEPHELINE SYENITE AND
BAUXITE AND BAUXITIC CLAY SAMPLES FROM ARKANSAS[1]

Element	Nepheline syenite (3 samples) (wt.%)	Bauxite, all types (14 samples) (wt.%)	Bauxite deposits			
			Type 1 (3 samples) (wt.%)	Type 2 (7 samples) (wt.%)	Type 3 (3 samples) (wt.%)	Type 4 (1 sample) (wt.%)
Ti	0.51[2]	1.06	0.73	1.1	1.0	1.8
Zr	0.050	0.13	0.12	0.12	0.13	0.2
Mn	0.097	0.12	0.083	0.17	0.040	0.05
Ca	1.01[2]	0.12	0.033	0.11	0.20	0.2
Nb	0.013	0.050	0.050	0.050	0.040	0.08
Mg	0.38[2]	0.037	0.0053	0.050	0.037	0.09
Sr	0.027	0.019	0.0035	0.010	0.053	0.03
Ba	0.071	0.017	0.0013	0.0030	0.060	0.04
Y	0.013	0.015	0.0057	0.020	0.012	0.02
Cr	0.0001	0.011	0.0043	0.0089	0.018	0.02
La	0.030	0.010	0.012	0.0096	0.0067	0.02
V	0.0047	0.0092	0.0060	0.0077	0.016	0.01
Ga	0.0020	0.0086	0.0063	0.0060	0.0093	0.008
B	—	0.0031	—	0.0029	0.0070	0.002
Mo	0.00057	0.0018	0.0010	0.0016	0.0030	0.002
Cu	0.00017	0.0014	0.0013	0.0019	0.00067	0.001
Sc	0.00033	0.00069	0.00053	0.00074	0.00077	0.0006
Pb	0.00067	0.00067	0.00033	0.00067	0.00090	0.001
Ni	—	0.00064	—	—	0.0030	—
Be	0.00017	0.00022	—	0.000057	0.00090	—
Co	—	0.00014	—	—	0.00067	—

[1] Analyst K. J. Murata; [2] Calculated from chemical analyses.
Looked for but not found: in nepheline syenites: As, Sb, Bi, In, Ti, Sn, Ag, Zn, Cd; in bauxites and bauxitic clays: As, Sb, Bi, In, Ti, Sn, Ge, Ta, P.

riched in the allite zone of the swamp bauxites. Disintegration of goethite at negative Eh values may cause Co and Ni removal. Manganese, too, readily dissolves in a reducing environment. In the allite zone Cr and Zr are associated both with relic minerals and neomineralization (Table XXXIX).

Not all bauxite deposits show such straightforward behaviour of trace elements. KHALIGHI (1968) investigated the distribution of Mn, Ni, Cr and P in a greater number of bauxite and laterite profiles on basalts and charnockites in India. He observed that there are no simple relationships between these elements and anyone of the major elements, but the values of the trace elements may fluctuate widely in various samples (Table XL).

Both removal and enrichment of Cr and P are found in various bauxites. In some bauxites Ni may be enriched readily, but it may have completely migrated

TABLE XXXIV

RELATIVE CONCENTRATION OR DEPLETION OF TRACE ELEMENTS IN BAUXITE AND BAUXITIC CLAY AS COMPARED TO NEPHELINE SYENITE (after GORDON et al., 1958)

Elements	Ratios of concentration				
	in bauxite all types	in bauxite deposits			
		Type 1	Type 2	Type 3	Type 4
Cr	100	40	90	180	200
Cu	8	8	11	4	6
Ga	4.3	3.2	3.0	4.7	4
Nb	3.8	3.8	3.8	3.1	6
		[3.0]			
Mo	3.2	1.8	2.8	5.3	4
[Al]	[2.7]	[2.7]	[2.4]	[3.0?]	
Zr	2.6	2.4	2.4	2.6	4
Ti	2.1	1.4	2.1	2.0	[2.9]
Sc	2.1	1.6	2.2	2.3	2
V	2.0	1.3	1.6	3.4	4
Be	1.3	—	0.3	5.3	—
Mn	1.2	0.9	1.8	0.4	0.5
Y	1.2	0.4	1.5	0.9	1.5
Pb	1.0	0.5	1.0	1.3	1.5
Sr	0.7	0.1	0.4	2.0	1.0
La	0.3	0.4	0.3	0.2	0.7
Ba	0.2	0.02	0.04	0.8	0.6
Ca	0.1	0.03	0.1	0.2	0.2
Mg	0.1	0.01	0.1	0.1	0.2

(Left-margin labels: IV > ; Concentrated < Al ; Depleted)

from other bauxites. The most widely fluctuating element is manganese. Both strong leaching and enrichment of Ni and Mn occur in the topmost part of various profiles. More data are required to characterize the behaviour of these elements during lateritic weathering.

Trace elements in karst bauxites

In karst bauxites the distribution of trace elements is governed by the nature of source rocks, and the degree of drainage.

According to SCHROLL et al. (1963) the elements Cr, Ni, Be and B followed the pattern of iron enrichment in bauxites of Austria. As trace elements of these bauxites originate from source rocks to a large extent, sufficient knowledge of geochemical bauxite composition might provide a means of classifying the

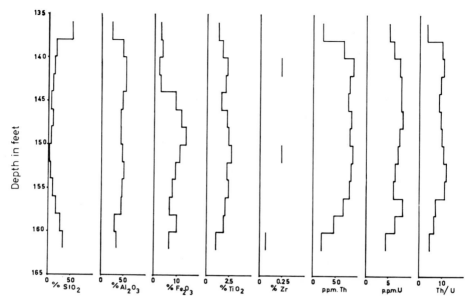

Fig.77. Chemical distribution of several major and trace elements in bauxites on nepheline syenite in Saline County, Ark., U.S.A. Bores NE 1/4, SE 1/4, sect. Tp. 25, R 14 W. (After ADAMS and RICHARDSON, 1960.)

TABLE XXXV

COMPOSITION OF ANDESITE AT MUNGGU BELIAN, SEMATAN (after WOLFENDEN, 1965)

Major elements	(%)	Minor elements	(p.p.m.)
SiO_2	52.49	Cr	75
TiO_2	1.82	Co	65
Al_2O_3	17.70	Ga	16
Fe_2O_3	5.84	Mn	700
FeO	2.16	Ni	140
MnO	0.10	Sr	720
MgO	4.16	V	235
CaO	7.39	Y	100
Na_2O	3.98	Zr	240
K_2O	1.00		
H_2O^+	1.83		
H_2O^-	1.40		
P_2O_5	0.83		

Major elements determined by the Mineral Resources Division, Overseas Geological Surveys, London.

TABLE XXXVI

PARTIAL ANALYSES OF HILL BAUXITE AT MUNGGU BELIAN, SEMATAN (after WOLFENDEN, 1965)

Specimen no.:	S7271	S7272	S7275	S7276	S7277	S7278	S7279	S7280	Average
Major elements (%)									
Al_2O_3	54.20	55.62	56.44	51.89	54.57	55.18	54.49	55.48	54.73
Combined SiO_2	1.02	1.61	1.14	1.87	2.46	0.88	0.78	0.97	1.34
Free SiO_2	0.50	0.37	0.21	0.30	0.36	0.37	0.45	0.23	0.35
Fe_2O_3	9.34	9.05	8.57	11.57	8.74	9.38	10.10	8.64	9.42
FeO	0.47	0.44	0.57	0.55	0.40	0.44	0.44	0.54	0.48
TiO_2	2.52	2.42	2.25	3.21	2.30	2.49	2.47	2.08	2.47
P_2O_5	0.32	0.17	0.34	0.29	0.21	0.31	0.33	0.32	0.28
Loss on ignition	29.45	29.19	29.99	28.34	29.31	29.33	29.40	30.10	29.39
H_2O ($-105\,°C$)	0.70	0.42	0.51	0.65	0.54	0.45	0.63	0.54	0.55
Total	98.52	99.29	100.02	98.67	98.89	98.83	99.09	98.90	99.01
Minor elements (p.p.m.)									
Cr	260	210	250	320	290	250	280	250	260
Co	39	22	20	10	—	33	24	20	21
Ga	28	30	24	22	25	37	27	22	27
Mn	290	370	320	370	220	290	320	300	310
Ni	47	45	90	39	45	48	50	43	50
Sr	—	170	85	190	140	45	50	130	~100
V	270	200	235	300	228	235	235	200	240
Y	—	—	—	—	—	—	—	—	—
Zr	410	480	390	400	410	390	370	350	400

Major elements determined by the Mineral Resources Division, Overseas Geological Surveys, London; figures in p.p.m. represent concentrations in ignited rock; a dash (—) indicates: not detected.

source rocks. SCHROLL and SAUER (1968) published several ratios of elements of various source rocks and bauxites, e.g., Ti/Cr, Cr/Ni, Mo/V. He showed that bauxites can be characterized quite well by Cr/Ni ratios. On the basis of this ratio, the bauxites investigated by Schroll et al. appear to have originated from ultra-basic to basaltic rocks (Fig.78). The Ti/Cr or Mo/V ratios are less suitable for geochemical characterization of bauxite provinces.

MAKSIMOVIĆ (1968) demonstrated variation in trace element composition of bauxites to be a function of source rocks of various bauxite deposits in Yugoslavia. In karstbauxites of the western part of Herzegovina, Cr, Mo, Li, Ni, Zr, Y, La and Sr, for example, are less enriched but U and Th are more strongly enriched than in corresponding bauxites of the eastern province. The deposits in the east contain 2–5 times as much chromium as the bauxites in the west. These variations clearly reflect a different composition of the source material (Table XLI).

There is mobilization of both major elements and trace elements during strong diagenesis whereby trace elements may follow the pattern of enrichment or

TABLE XXXVII

PARTIAL ANALYSES OF SWAMP BAUXITE AT MUNGGU BELIAN, SEMATAN (after WOLFENDEN, 1965)

Specimen no.	S7268	S7269	S7270	S7273	S7281	S7282	Average
Major elements (%)							
Al_2O_3	49.95	52.50	49.56	54.90	47.72	54.83	51.58
Combined SiO_2	9.01	9.73	8.39	4.56	5.15	4.16	6.83
Free SiO_2	7.78	1.36	6.53	4.77	10.36	6.64	6.24
Fe_2O_3	2.32	5.25	4.05	3.22	4.33	1.60	3.46
FeO	0.41	0.21	0.35	0.22	0.70	0.21	0.35
TiO_2	2.76	2.59	2.95	1.93	2.28	1.83	2.39
P_2O_5	0.19	0.13	0.17	0.19	0.21	0.19	0.18
Loss on ignition	26.08	27.21	25.76	29.16	28.16	28.78	27.52
H_2O ($-105\,^{\circ}C$)	0.69	0.83	0.72	0.48	0.99	0.64	0.72
Total	99.19	99.81	98.48	99.43	99.90	98.88	99.27
Minor elements (p.p.m.)							
Cr	240	370	310	200	240	200	260
Co	9	10	5	7	13	7	8
Ga	47	42	42	38	40	34	40
Mn	140	120	320	120	130	70	150
Ni	40	44	20	17	31	12	27
Sr	120	130	220	170	110	160	150
V	185	235	170	210	280	130	200
Y	—	—	—	50	—	—	—
Zr	530	430	470	460	400	370	440

Major elements determined by the Mineral Resources Division, Overseas Geological Surveys, London; figures in p.p.m. represent concentrations in ignited rock; a dash (—) indicates: not detected.

migration from bauxites of specific major elements. In some places the elements Zr, Pb, U and Th are associated with clastic zircon minerals. Ti and Zr are built mainly into the anatase structure during neomineralization. Presumably Zr, V and Ga are accommodated by aluminium minerals, too. There is conformity in nickel and kaolinite enrichment to a large extent.

In bauxites of the Var, France, V and Zr concentrate with Ti and are likely to be built into the anatase structure (VALETON, 1965). This concentration progresses parallel to aluminium enrichment in the ferrallitic main zone of the bauxites (Fig.79).

However, in the siallitic top zone there is selective enrichment of V, Zr and Ti in iron concretions.

Concentrations of trace elements at the base of bauxite deposits

Many karst bauxite deposits in southern Europe are enriched with Mn, Ni and

TABLE XXXVIII

PARTIAL ANALYSES OF KAOLINITIC CLAY OF MUNGGU BELIAN, SEMATAN (after WOLFENDEN, 1965)

Specimen nos.:	S6108	S6539	S6546	S6547	S6557	Average
Major elements (%)						
Al_2O_3	29.51	35.00	31.17	24.47	29.89	30.01
SiO_2	34.72	30.50	24.90	26.53	33.40	30.01
Total Fe as Fe_2O_3	13.60	12.03	15.67	11.34	12.47	13.02
MgO	0.58	n.d.	n.d.	5.40	n.d.	—
CaO	0.06	n.d.	n.d.	3.60	n.d.	—
TiO_2	2.90	2.88	3.18	2.55	2.71	2.84
P_2O_5	0.23	0.28	1.90	2.33	0.65	1.08
Loss on ignition	17.28	18.76	22.54	21.87	19.93	20.08
Total	98.88	99.45	99.36	98.09	99.05	97.04[1]
Minor elements (p.p.m.)						
Cr	180	150	360[2]	160	160	160
Co	65	34	46	90	100	67
Ga	22	25	25	20	20	23
Mn	380	300	320	2200[2]	2200[2]	330
Ni	140	240	180	90	210	170
Sr	—	65	640[2]	—	180	—
V	270	250	340	310	230	280
Y	65	—	50	—	60	—
Zr	440	510	430	400	430	440

Major elements determined by the Mineral Resources Division, Overseas Geological Surveys, London; n.d. not determined; figures in p.p.m. represent concentrations in ignited rock; a dash (—) indicates: not detected.
[1] Low total as CaO and MgO in S6108 and S6547 not averaged; [2] these anomalously high values not considered in calculating averages.

frequently also with Co in the basal horizon. Such a horizon is mined as nickel ore in the bauxites of the Lokris region in Greece. In the neighbourhood of the lateritized ultrabasic belts of the sub-Pellagonic zone the bauxites of Neonkokkinon deposited on karstificated Jurassic limestone. At the base fine-grained material occurs with increasing intercalations of coarse clastic laterite bauxite towards the floor of the bauxites. There is an increase of the Ni and Co content in this direction (MOUSSOULOS, 1957). Ni and Co values are most abundant in clayey bauxites rich in iron at the basis and assay:

SiO_2	16.45%	NiO	2.67%
Fe_2O_3	53.50%	CoO	0.05%
Al_2O_3	13.45%	P_2O_5	0.09%
Cr_2O_3	1.38%	Mn_3O_4	2.25%
CaO	0.74%	CuO	0.07%
MgO	1.85%	SO_3	0.12%
		As	0.03%

TABLE XXXIX

CONCENTRATION RATIOS OF ELEMENTS IN BAUXITE AND KAOLINITIC CLAY AS COMPARED WITH ANDESITE SOURCE ROCK (after WOLFENDEN, 1965)

	Element	Ratios of concentration and depletion		
		hill bauxite	swamp bauxite	kaolinitic clay
Concentrated > Al	Cr	3.5	3.5	2.7
	Al	3.1	2.9	
	Zr	1.6	1.8	1.8
				1.7
	Ga	1.7	2.5	1.4
Concentrated < Al	Ti	1.4	1.3	1.6
	Fe	1.2	0.5	1.6
	V	1.0	0.9	1.2
	Ni	0.4	0.2	1.2
	Co	0.3	0.1	1.0
	P	0.3	0.2	1.3
Depleted	Mn	0.4	0.2	0.5
	Sr	0.14	0.2	—
	Si	0.03	0.3	0.6

TABLE XL

TRACE ELEMENTS (p.p.m.) IN CHARNOCKITE, BASALTS AND BAUXITES OF INDIA (after KHALIGHI, 1968)

Materials	Location	Mn	Cr	P	Ni
Charnockite	Kotagiri	3560	200	350	2100
Bauxites	Kotagiri	380–1970	40–100	10–1680	800–3600
Basalts	Udagiri	8040	150	750	3500
Saprolites	Udagiri	430–3030	190–370	180–740	100–1600
Bauxites	Udagiri	240–2690	220–980	160–620	100–2900
Iron crust	Udagiri	1360–13240	460–1530	340–1540	200–2100
Basalts	Mewasa/Gujerat	4680	140	300	1200
Saprolites	Mewasa/Gujerat	280–1820	50–110	150–700	200–1400
Bauxites	Mewasa/Gujerat	540–2120	40–160	90–250	300–500
Iron crust	Mewasa/Gujerat	520–1770	100–150	410–730	100–1000

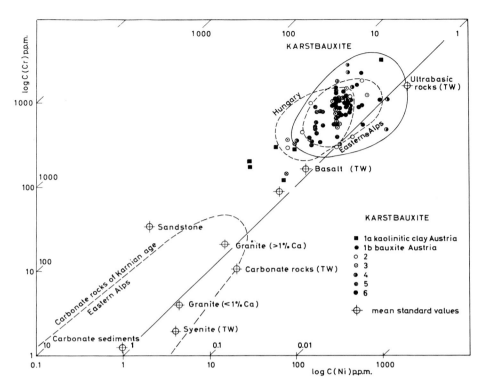

Fig.78. Binary correlation diagram of Cr/Ni in karst bauxites (analyses by SAUER, 1965) and carbonate rocks of the east Alpine Karn formation, with reference to mean values of different rocks obtained by TUREKIA and WEDEPOHL (1961).
1. Austria; *2.* Hungary; *3.* Yugoslavia; *4.* Greece; *5.* France, Spain; *6.* Jamaica. (After SCHROLL and SAUER, 1968.)

TABLE XLI

COMPARISON OF SOME TRACE ELEMENT CONTENTS IN BAUXITES FROM FOUR DEPOSITS OF WESTERN AND EASTERN HERZEGOVINA (after MAKSIMOVIĆ, 1968)

Deposit	*Number of samples*	*Average contents (p.p.m.)*							*Th/U*
		Cr	*Li*	*Ni*	*Y*	*La*	*U*	*Th*	
West Herzegovina:									
Studena Vrela	31	447	8.5	346	47	117	7.6	40.0	5.2
Mratnjača	10	568	4.5	333	47	143	7.0	39.2	5.6
East Herzegovina:									
Dabrica	13	1150	71	371	168	273	2.7	34.6	12.8
Gornji Brštanik	4	1600	74	500	272	205	3.0	39.8	13.2

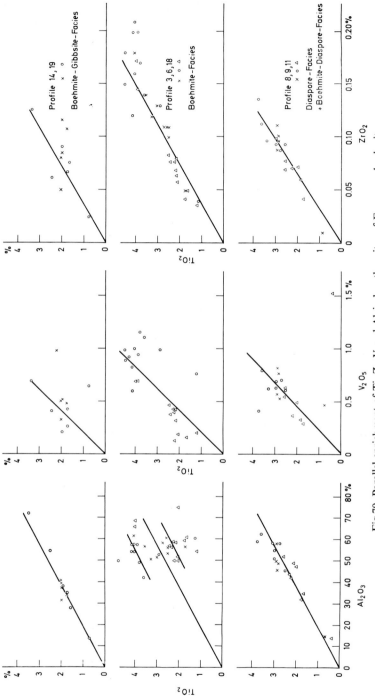

Fig.79. Parallel enrichment of Ti, Zr, V and Al in karstbauxites of France: *a.* boehmite–gibbsite facies of Les Baux; *b.* boehmite facies, Var; *c.* diaspore and boehmite facies of Aude and Ariège. In the main zone of the bauxite these elements are probably built into the structures of aluminium and titanium minerals. (After VALETON, 1965.)

Enrichment is caused by dissolution of nickel and cobalt in the upper part of the profile with irregular impregnation of the lowest centimetres (usually several centimetres; in rare cases up to 3 m) during late diagenesis. Ni and Mn enrichments at the base of bauxite were also observed in other Cretaceous bauxites of Greece (BARDOSSY and MACK, 1967), France (VALETON, 1966; DUROZOY et al., 1966) and Hungary. They often form irregular black crusts on the karst surface several centimetres thick (Plate VIII, 32).

The nickel and manganese minerals known to occur are:

garnierite $Ni_6[(OH)_8/Si_4O_{10}]$

takovite $5 Ni(OH)_2 \cdot 4 Al(OH)_3 \cdot 4 H_2O$

lithioporite $(Al_2Li)(OH)_2MnO_2$

todorokite $(H_2O) \leqslant 2 (Mn \ldots) \leqslant 8 (O, OH)_{16}$

WATER

In practice the water content of bauxites with traces of CO_2 and SO_4 is determined as ignition loss.

The volatile components of major bauxite minerals are (Table XLII): H_2O^+, CO_2 and SO_4 (originating from S^{2-}).

The aluminium minerals contain most of the water of the bauxites. Several percent excess water not attributable to bauxite minerals occurs in holocrystalline bauxites of India and France. The same phenomenon was observed in bauxites of Surinam by G. J. J. Aleva (personal communication). However, there is no solution yet to the problem of coordination of excess water.

From laterite bauxites of India ROY CHOWDHURY et al. (1964) report that Al_2O_3 content, hence bauxite quality, may be inferred from ignition loss under certain conditions, e.g., if the clay-mineral content is negligible, and if hematite and gibbsite are the only iron and aluminium minerals, respectively.

TABLE XLII

VOLATILE COMPONENTS (%) OF MAJOR BAUXITE MINERALS

Minerals	H_2O^+	Al_2O_3	SiO_2	Fe_2O_3	FeO	CO_2	S
Gibbsite	34.6	65.4	–	–	–	–	–
Boehmite	15.0	85.0	–	–	–	–	–
Diaspore	15.0	85.0	–	–	–	–	–
Kaolinite group	14.0	39.5	46.5	–	–	–	–
Hematite	–	–	–	100.0	–	–	–
Goethite	10.1	–	–	89.9	–	–	–
Siderite	–	–	–	–	62.1	37.9	–
Pyrite	–	–	–	–	46.6	–	53.4

Chapter 7

PROBLEMS AWAITING SOLUTION

The previous chapters are a compilation of the known geology, mineralogy and geochemistry of bauxite deposits, and the author tried to synthesize the geological and geochemical laws underlying bauxite genesis. Unfortunately, for obvious linguistic reasons, important Russian publications have not been given the same weight as English, French and German literature. Our present-day knowledge is far from complete, and a great number of problems concerning the geology, mineralogy and geochemistry of bauxite deposits still await solution.

In this context only a few important problems which may be solved in the near future are referred to.

The questions concerning *bauxites on igneous and metamorphic rocks*, particularly in Africa, are:

age of the deposits

lower boundary and texture of the bauxite bodies

distribution and interfingering of gibbsite and boehmite facies

characterization of various neomineralization textures in gibbsite and boehmite facies

conditions of development of primary hematite or goethite facies

geochemical distribution of trace elements as function of primary mineral facies

characterization of polygenetic facies alteration e.g., by precipitation of secondary iron minerals such as maghemite and siderite

geochemistry: (*a*) in an oxidizing environment; (*b*) in a reducing environment.

The questions concerning *bauxites on karst* concern:

origin and composition of the source material

extent of superimposed diagenetic mineral formations

role of diaspore and iron minerals during diagenesis

characterization of neomineralization textures, depending on facies province, time, kind and extent of trace element redistribution.

In spite of the size of *bauxite deposits on clastic sediments*, very little research has been carried out and important problems are:

mechanism of aluminium supply to sedimentary basins

sedimentary environment and fossils or traces of fossils

facies paleogeography of gibbsitic and boehmitic bauxites
type and distribution of neomineralization textures
geochemical laws of distribution patterns and accommodation of trace
elements
significance of phosphorus.

Only after a sufficiently accurate solution of all these problems will it be possible to assess the time-space constellation governing optimum bauxite formation and to characterize bauxite formation during the earth's history at an advanced level. Also the relationship of the distributive province and paragenesis on the one hand and geochemistry on the other will become clearer after such an assessment. A better understanding of geochemical behaviour of trace elements, however, may be achieved only in the distant future.

References

ABBOT, A. T., 1958. Occurrence of gibbsite on the island of Kauai, Hawaiian Islands. *Econ. Geol.*, 53: 842–853.

ADAMS, J. A. S. and RICHARDSON, K. A., 1960. Thorium, uranium and zirconium concentrations in bauxite. *Econ. Geol.*, 55: 1653–1675.

AHMAD, N., 1966. Genesis, mineralogy and related properties of West Indian soils, 1. Bauxitic soils of Jamaica. *Soil. Sci. Soc. Am., Proc.*, 30: 719–722.

ALEVA, G. J. J., 1965. The buried bauxite deposit of Onverdacht, Surinam, South America. *Geol. Mijnbouw*, 44: 45–58.

ALLEN, V. T., 1935. Mineral composition and origin of Missouri flint and diaspore clays. *Mem. Geol. Surv., Bien. Rept., App.IV*, 58: 5–24.

ALLEN, V. T., 1952. Petrographic relations in some typical bauxite and diaspore deposits. *Bull. Geol. Soc. Am.*, 63(7): 649–688.

ALLEN, V. T., 1954. Relation of porosity and permeability to the origin of diaspore clay. *Clays Clay Minerals*, 3: 389–401.

ALLEN, V. T. and SHERMAN, G. D., 1965. Genesis of Hawaiian bauxite. *Econ. Geol.*, 60(1): 89–99.

ALTSCHULER, Z. S., DWORNIK, E. J. and KRAMER, H., 1963. Transformation of montmorillonite to kaolinite during weathering. *Science*, 141: 148–152.

ALWAR, M. A. A., 1962. Bauxite deposits in the Karagpur Hills, Monghyr district, Bihar. *Indian Minerals*, 16(1): 30–33.

ARKHANGELSKY, A. D., 1933. Origin of the bauxites of the U.S.S.R. Bull. Soc. Nat. Moscow, Sect. Geol. 11: 434–436.

A.S.T.M., 1963. *Index (Inorganic) to the Powder Diffraction*. Am. Soc. Testing Materials. Philadelphia, Pa.

AUBERT, G., 1965. Classification des sols. Tableaux des classes, sous-classes, groupes et sous-groupes de sols utilisés par la section de Pédologie de l'ORSTOM. *ORSTOM Cahiers Pedol.*, 1965(3): 269–288.

BALKAY and BARDOSSY, G., 1967. Lateritesedési részfolyamat vizsgálatok Guineai lateriteken. *Földt. Közl.*, 97: 91–110.

BARDOSSY, G., 1954. Melanterit a scozi bauxitban. *Földt. Közl.*, 84(3): 217–219.

BARDOSSY, G., 1957. Csigamaradvány a nagykovácsi agyagos bauxitból. *Földt. Közl.*, 87: p.454.

BARDOSSY, G., 1958. Geochemistry of Hungarian bauxites. *Acta Geol. Acad. Sci. Hung.*, 5: 103–155; 255–285.

BARDOSSY, G., 1961. *A Magyar Bauxit Geokémiai Vizsgálata*. Muszaki Könyvkiadó, Budapest, 233 pp.

BARDOSSY, G., 1963. Die Entwicklung der Bauxitgeologie seit 1950. *Symp. Bauxites, Zagreb, 1963*, 1: 31–50.

BARDOSSY, G., 1965. Quelques problèmes de la minéralogie des bauxites. *Comm. Soc. Franç. Minéral.-Crist.*

BARDOSSY, G., 1966. *Bibliographie des Travaux concernant les Bauxites publiés en français, anglais, russe et allemand*. ICOBA, Paris, 51 pp.

BARDOSSY, G. and BARDOSSY, L., 1954. Contribution to the geochemistry of titanium. *Acta Geol. Acad. Sci. Hung.*, 2: 191–203.

BARDOSSY, G. and MACK, E., 1967. Zur Kenntnis der Bauxite des Parnass-Kiona-Gebirges. *Mineral. Deposita*, 2: 334–348.

BARNABAS, K., 1961. Stratigraficeskoe polozenie mestoro zdenij melovich boksitov Vengrii. *M.A.F.J. Evkönyve (Budapest)* 49: 1005–1016.

BATES, T. F., 1962. Halloysite and gibbsite formation in Hawaii. *Clays Clay Minerals, Proc. Natl. Conf., Clays Clay Minerals*, 9(1960): 315–328.

BAUXITSYMPOSIUM, 1963. *Symposium sur les Bauxites, Oxydes et Hydroxides d'Aluminium*, 1–3. *Bauxitsymposium*, Zagreb, 553 pp.

BENESLAVSKY. S. J., 1959. Chemical and mineralogical composition of bauxites and some problems concerning the genesis of their minerals. *Acta Geol. Acad. Sci. Hung.*, 6(1/2): 55–64.

BENESLAVSKY, S. J., 1963. *Minéralogie des Bauxites (Critères d'Evaluation de la Quantité et des Propriétés technologiques des Minerais bauxitiques d'après leur Composition)*. Gosgeolve Khizdat, Moscow, 170 pp. (Traduction No. 4706 B.R.G.M., Paris).

BERTHIER, P., 1821. Analyse de l'alumine hydratée des Baux. *Ann. Mines*, 6: 531–534.

BETECHTIN, 1964. *Lehrbuch der speziellen Mineralogie*. Deutscher Verlag Grundstoff-Industrie, Leipzig, 679 pp.

BEZJAK, A. and JELENIĆ, J., 1963. The crystal structures of boehmite and bayerite. *Symp.* ICOBA, *Zagreb, 1963*: 105–112.

BIEDERMANN, G. and SCHINDLER, P., 1957. On the solubility of precipitated iron (III) hydroxide. *Acta Chem. Scand.*, 11: 731–740.

BIELFELDT, K., 1968. Entwicklungsrichtungen der Aluminium-Industrie, insbesondere im Hinblick auf die Standortwahl. *Erzmetall*, 21: 547–598.

BLEACKLEY, D., 1960. The bauxites and laterites of British Guiana. *Geol. Surv. Brit. Guiana*, 34: 1–156.

BOLGER, R. C. and WEITZ, J. H., 1951. Mineralogy and nomenclature of the Mercer fireclay in north-central Pennsylvania. *Proc. Penn. Acad. Sci.*, 25: 124–130.

BOLGER, R. C. and WEITZ, J. H., 1952. Mineralogy and origin of the Mercer fireclay of north-central Pennsylvania. *A.I.M.E. Symp. Vol.—Problems of clay and laterite genesis*: 81–93.

BONIFAS, M., 1959. Contribution à l'étude géochimique de l'altération latéritique. *Mém. Serv. Carte Géol. Alsace Lorraine*, 17: 1–159.

BONIFAS, M. and LEGOUX, P., 1957. Présence de maghémite massive dans des produits d'altération latéritique. *Bull. Serv. Carte Géol. Alsace Lorraine*, 10: 7–9.

BONTE, A., 1965. Sur la formation en deux temps des bauxites. *Compt. Rend.*, 260: 5076–5077.

BRIDGE, J., 1950. Bauxite deposits of the southeastern United States. In: *Symposium on Mineral Resources of the southeastern United States*. Tennessee Univ. Press, Knoxville, Tenn., pp.107–201.

BRINDLEY, G. W., 1961. The reaction series gibbsite–chi-alumina–kappa-alumina–corund. *Am. Mineralogist*, 46: 771–785; 1187–1190.

BROWN, G., 1953. The dioctahedral analogue of vermiculite. *Clay Minerals Bull.*, 2(10): 64–69.

BUCHINSKY, G. J., 1963. Types of karst bauxite deposits and their genesis. *Symp. ICOBA, Zagreb, 1963*, 1: 93–105.

BUCHINSKY, G. J., 1966. Progress in the study of bauxite genesis for the last ten years (1955–1965). In: *The Genesis of Bauxites*. Geol. Inst. Acad. Sci., U.S.S.R., Moscow, pp.3–30 (in Russian).

BUNSEN, R., 1852. Darstellung des Magnesiums auf elektrolytischem Wege. *Ann. Chem.*, 82: 137–145.

BUTTERLIN, J., 1958. A propos de l'origine des bauxites des régions tropicales calcaires. *Compt. Rend. Soc. Géol. France*, 5: 121–123.

CAILLÈRE, S., 1962. Boehmite et diaspore ferrifères dans un bauxite de Pèreille. *Compt. Rend.*, 254: 137–139.

CAILLÈRE, S. and POBEGUIN, T., 1965. Considérations générales sur la composition minéralogique et la genèse des bauxites du Midi de la France. *Mém. Museum Natl. Hist. Nat. (Paris), Sér. C*, 12(4): 125–212.

CAILLÈRE, S., HÉNIN, S. and POBEGUIN, T., 1962. Présence d'un nouveau type de chlorite dans les "bauxites" de St. Paul de Fenouillet, Pyrénées Orientales. *Compt. Rend.*, 254: 1657–1658.

CAPDECOMME, L., 1952. Études minéralogiques des gîtes de phosphates alumineux de la région de Thiès (Senegal). Origine de gisements de phosphates de Chaux. *Congr. Géol. Intern., Compt. Rend., 19e, Algiers, 1952*: 103–117.

CELET, P., 1962. Contribution à l'étude géologique du Parnasse-Kiona et d'une partie des régions méridionales de la Grèce continentale. *Ann. Géol. Pays Helléniques*, 13: 446 pp.

CHOUBERT, B., 1965. Nos connaissances sur la géologie de la Guyane française. *Bull. Soc. Géol. France, Sér. 7*, 1: 129–135.

CHOWDHURY, A. N., CHAKRABERTY, S. C. and BOSE, B. B., 1965. Geochemistry of gallium in bauxite from India. *Econ. Geol.*, 60: 1057.

CLARKE, O. M., 1966. The formation of bauxite on karst topographic in Eufalia district, Alabama, and Jamaica, West India. *Econ. Geol.*, 61: 903–916.

COMBES, P. J., 1965. Dissolution karstique sous une couche de bauxite. Remarques sur l'origine des gisements en poches. *Compt. Rend. Soc. Somm. Géol. France*, 4: 123–124.

COMBES, P. J. and REY, J., 1963. Découverte de bauxites intraurgoniennes de la région de Durban-sur Ariège (Ariège). *Compt. Rend. Somm. Soc. Géol. France*, 9: 318–320.

COQUAND, M. H., 1871. Sur les bauxites de la chaîne des Alpines (Bouche du Rhône). *Bull. Soc. Géol. France, 2,28*: 98–115.

CORRENS, C. W., 1963. Experiments on the decomposition of silicates and discussion of chemical weathering. *Clays Clay Minerals Symp. Publ.*, 1963: 443–459.

CORRENS, C. W. and VON ENGELHARDT, W., 1938. Neue Untersuchungen über die Verwitterung des Kalifeldspates. *Chem. Erde*, 12: 1–22.

CORRENS, C. W. and VON ENGELHARDT, W., 1941. Röntgenographische Untersuchungen über den Mineralbestand sedimentärer Eisenerze. *Nachr. Akad. Wiss. Göttingen, Math. Phys. Kl.*, 131–137.

DE KIMPE, C., GASTUCHE, M. and BRINDLEY, G. W., 1961. Ionic coordination in alumo-silic gels in relation to clay mineral formation. *Am. Mineralogist*, 46: 1370–1381.

DE LAPPARENT, J., 1930. *Les Bauxites de la France méridionale.* Imprimerie Nationale, Paris, 186 pp.

DELVIGNE, J., 1965. *Pédogenèse en Zone tropicale. La Formation des Minéraux secondaires en Milieu ferrallitique.* ORSTOM, Paris 1965, 177 pp.

DEMANGERON, P., 1965. Sur la présence et la signification probable de minéraux du Massif central dans les bauxites de l'isthme durancien. *Compt. Rend.*, 261: 2685–2688.

DENIZOT, G., 1934. Description des massifs de Marseille, Veyre et de Puget. *Ann. Musée Hist. Nat., Marseille, Mém.*, 26(5): 236.

DE WEISSE, J. G., 1948. Les bauxites de l'Europe Centrale. *Mém. Soc. Vaudoise Sci. Nat.*, 58(9): 1–162.

DE WEISSE, J. G., 1968. Geographische Lage und wirtschaftliche Aspekte der bedeutendsten Bauxitlagerstätten der Erde. *Aluminium*, 44: 579–581.

DIEULAFAIT, L., 1881. Les bauxites, leurs âges, leurs origine, diffusion complète du titane et du vanadium dans les roches de la formation primordiale. *Compt. Rend.*, 93: 804–807.

DOEVE, G. and GROENEVELD MEIJER, W. O. J., 1963. Bauxite deposits of British Guiana and Surinam in relation to underlying unconsolidated sediments suggesting two-step origin. *Econ. Geol.*, 58(7): 1160–1162.

DUBOUL-RAXAVET, C. and PÉRINET, G., 1960. Sur la composition minéralogique des bauxites des Alpines. *Compt. Rend. Congr. Soc. Sav., 84e, Dijon, 1960*: 397–404.

DUCHAUFOUR, P., 1965. *Précis de Pédologie.* Masson, Paris, 481 pp.

DUFFIN, W. J. and GOODYEAR, J., 1960. A thermal X-ray investigation of scarbroite. *Mineral Mag.*, 32: 353–362.

DUNLOP, J. C., BERGQUIST, H. R., CRAIG, L. C. and OVERSTREET, E. F., 1965. Bauxite deposits of Tennessee. *U.S. Geol. Surv., Bull.*, 1199-L: 37 pp.

DUROZOY, G., HAUTE, H. and JACOB, C., 1966. Présence d'hydroxide de nickel naturel dans le gisement de bauxite de Codouls/Var. *Compt. Rend.*, 263(7): 625–626.

ERHART, H., 1956. *La Genèse des Sols en tant que Phénomène géologique.* Masson, Paris, 90 pp.

ERHART, H., 1965. Le témoignage paléoclimatique de quelques formations paléopédiques dans leur rapport avec la sédimentologie. *Geol. Rundschau*, 54: 15–24.

ERNST, L., 1968. Die Hüttenaluminiumproduktion der Welt im Jahre 1967. *Wirtschaft*, 44: 261–265.

ERWIN, G. and OSBORN, E. F., 1959. The system Al_2O_3–H_2O. *J. Geol.*, 59: 381–394.

EVANS, H. J., 1959. The geology and exploration of the Cape York Peninsular bauxite deposits in northern Queensland. *Chem. Eng. Mining, Rev.*, 51(11): 48–56.

EWING, F. J., 1935. The crystal structure of diaspore. *J. Chem. Phys.*, 3: 203–207.

EYLES, V. A., 1958. A phosphatic band underlying bauxite deposits in Jamaica. *Nature*, 182(4646): 1367–1368.

FEITKNECHT, W. and MICHAELIS, W., 1962. Über die Hydrolyse von Eisen (III) Perchloratlösungen. *Helv. Chim. Acta*, 45: 212–224.

FILIPOVSKI, GJ. and CIRIC, M., 1963. Zemljista Jugoslavije. *Yugoslav Soc. Soil Sci.*, 9: 1–500.

FISCHER, E. C., 1955. Annotated bibliography of the bauxite deposits of the world. *U.S. Geol. Surv. Bull.*, 999: 221 pp.

FOOSE, R. M., 1944. High-alumina clays of Pennsylvania. *Econ. Geol.*, 39: 557–577.

FOOTE, R. B., 1876. The iron clay of South India. *Mem. Indian Geol. Surv.*, 12: 217.

FOX, C. S., 1923. The bauxite and aluminous laterite occurrences of India. *Mem. Geol. Surv. India*, 49: 287 pp.

FOX, C. S., 1927. *Bauxite*. Crosby-Lockwood, London, 312 pp.

GARRELS, R. M. and CHRIST, G. L., 1965. *Solutions, Minerals and Equilibria*. Harper and Row, New York, N.Y., 450 pp.

GINSBERG, H., 1962. *Aluminium. (Die metallischen Rohstoffe, 15)*. Enke, Stuttgart, 135 pp.

GINSBERG, H., HÜTTIG, W. and STIEHL, H., 1962. Beiträge zum System H_2O–Al_2O_3, 2. *Z. Anorg. Allgem. Chem.*, 318: 238–256.

GLADKOVSKY, A. K. and USHATINSKY, J. N., 1963. Genesis and alteration of aluminous minerals in bauxite. *Symp. Bauxites, Zagreb, 1963*: 153–170.

GOLDICH, S. S., 1938. A study in rock-weathering. *J. Geol.*, 46: 17–58.

GOLDICH, S. S. and BERGQUIST, H. R., 1947. Aluminous lateritic soil of the Sierra of Bahoruco area (Dominican Republic). *U.S. Geol. Surv. Bull.*, 953C: 53–84.

GOLDICH, S. S. and BERGQUIST, H. R., 1948. Aluminous lateritic soil of the republic of Haiti, West Indies. *U.S. Geol. Surv. Bull.*, 954C: 109 pp.

GOLDSCHMIDT, V. M., 1937. The principles of distribution of chemical elements in minerals and rocks. *J. Chem. Soc.*, 139: 655–675.

GORDON, M., TRACEY, J. I. and ELLIS, M. W., 1958. Geology of the Arkansas bauxite region. *U.S. Geol. Surv., Profess. Papers*, 299: 268 pp.

GROSS, S. and HELLER, L., 1963. A natural occurrence of bayerite. *Mineral. Mag.*, 33: 723–724.

GROSSER, G., 1932a. Der Verwitterungsverlauf eines Basaltes aus der Oberlausitz. *Fortschr. Mineral. Krist. Petrog.*, 16: 327–328.

GROSSER, G., 1932b. Die Verfahren zur Berechnung und graphischen Darstellung der chemischen Gesteinsverwitterung, 1. Mitteilung über den Basalt des Wacheberges bei Taubenheim an der Spree. *Chem. Erde*, 7: 130–152.

GROSSER, G., 1935. Die Veränderungen im Chemismus der Eruptivgesteine durch die Verwitterung. *Chem. Erde*, 11: 73–216.

GRUBB, P. L. C., 1963. Critical factors in the genesis, extent and grade of some residual bauxite deposits. *Econ. Geol.*, 58: 1267–1277.

GRUBIC, A., 1964. Les bauxites de la province dinarique. *Bull. Soc. Géol. France*, 7(6): 382–388.

HABER, F., 1925. Über Hydroxyde des Al und Te^{3+}. *Naturwissenschaften*, 13: 1007–1012.

HABERFELLNER, E., 1951. Zur Genesis der Bauxite in den Alpen und Dinariden. *Berg Hüttenmänn, Monatsh.*, 96: 62–69.

HARDEN, G. and BATESON, J. H., 1963. A geochemical approach to the problem of bauxite genesis in British Guiana. *Econ. Geol.*, 58: 1301–1308.

HARDER, E. C., 1952. Examples of bauxite deposits illustrating variation in origin. Problems of clay and laterite genesis. *Am. Inst. Mining Met. Eng.*, 1952: 35–64.

HARDER, H., 1965. Experimente zur "Ausfällung" der Kieselsäure. *Geochim. Cosmochim. Acta*, 29: 429–442.

HARRASSOWITZ, H., 1926. Laterit. *Fortschr. Geol. Palaeontol.*, 4: 253–566.

HARTMANN, J. A., 1955. Origin of heavy minerals in Jamaican bauxite. *Econ. Geol.*, 50(7): 738–747.

HATHAWAY, J. C. and SCHLANGER, S. O., 1965. Nordstrandite ($Al_2O_3 \cdot 3H_2O$) from Guam. *Am. Mineralogist*, 50: 1029–1037.

HAUSCHILD, U., 1964. Alkalimetallionen freier Hydrargillit, $Al(OH)_3$. *Naturwissenschaften*, 51: 238–239.

HAZRA, P. C. and RAY, D. K., 1962. A short note on seismic phenomena in India and their relation to tectonics. *Indian Minerals*, 16: 241–246.

HENDRICKS, D. M., WITTIG, L. D., and JACKSON, M. D., 1967. Clay mineralogy of andesite saprolite. *Clays Clay Minerals, Proc. Natl. Conf. Clays Clay Minerals*, 27: 395–407.

HEYDEMANN, A., 1966. Über die chemische Verwitterung von Tonmineralen (experimentelle Untersuchungen). *Geochim. Cosmochim. Acta*, 30(10): 995–1035.

HILL, V. G., 1955. The mineralogy and genesis of bauxite deposits of Jamaica. *Am. Mineralogist*, 40(7/8): 676–688.

HILLER, J. E., 1964. *Kristallchemie des FeOOH*. Deutsche Mineralogische Gesellschaft, Wiesbaden, 1 p.

HOSE, H. R., 1950. Geology and mineral resources of Jamaica. *Colonial Geol. Mineral Resources (Gt. Brit.)*, 1: 11–36.

HOSE, H. R., 1963. Bauxite Mineralogy. *Extract. Met. Aluminium*, 1: 3–20.

HUDDLE, J. W. and PATTERSON, S. H., 1961. Origin of Pennsylvanian underclay and related seat rocks. *Geol. Soc. Am., Bull.*, 72: 1643–1660.

KALUGIN, A. S., 1967. On the role of volcanicity and reefs in bauxite formation of geosynclinal areas. *Lithol. Mineral Res.*, 1: 3–22 (in Russian).

KARSULIN, M., 1963. Das Mineral 2 $Al_2(OH)_6 \cdot 1H_2O$-"Tućanit". *Symp. Bauxite, Zagreb*, 2: 37–46.

KATSURA, T., KUSHIRO, I., AKIMOTO, S. J., WALKER, J. L. and SHERMAN, G. D., 1962. Titanomagnetite and titanomaghemite in a Hawaiian soil. *J. Sediment. Petrol.*, 32: 299–308.

KELLER, W. D., 1952. Observation on the origin of Missouri high-alumina clays. *A. I. M.E. Symp. Vol.—Problems of clay and laterite genesis*, 1952: 115–135.

KELLER, W. D., 1964. The origin of high-aluminia clay minerals, a review. *Clay Clay Minerals*, 19: 129–151.

KELLER, W. D., WESTCOTT, J. F. and BLADSOE, A. O., 1954. The origin of Missouri fire clays. *Clay Clay Minerals*, 2: 7–46.

KELLOGG, C. E., 1949. Preliminary suggestions for the classification and nomenclature of great soil groups in tropical and equatorial regions. *Comm. Bur. Soil Sci., Tech. Commun.*, 46: 76–85.

KENNEDY, G. C., 1959. Phase relations in the system Al_2O_3–H_2O at high temperatures and pressures. *Am. J. Sci.*, 257: 563–573.

KESSMANN, J., 1966. Zur hydrothermalen Synthese von Brookit. *Z. Anorg. Allgem. Chem.*, 346: 30–43.

KHALIGHI, M., 1968. *Zur Untersuchung der Spurenelemente in den Indischen Bauxiten, Lateriten und deren Ausgangsgesteinen Basalt und Charnockit*. Dipl. Arbeit, Hamburg, 76 pp. (unpublished).

KISPATIC, M., 1912. Bauxite des kroatischen Karstes und ihre Entstehung. *Neues Jahrb. Mineral.*, 34: 513–522.

KISS, J. and VÖRÖS, J., 1965. La bauxite lignitifère du mont Bagolyhegy (Gánt) et le mécanisme de la sedimentation de la bauxite. *Ann. Univ. Sci. Budapest., Sect. Geol.*, 8: 67–90.

KÖHLER, A., 1955. Ein Vorkommen von Carnotit im Bauxit von Unterlaussa. *Jahrb. Oberösterr. Musealvereins*, 100: 359–360.

KOMLOSSY, G., 1967. Contributions à la connaissance de la génèse des bauxites hongroises. *Acta. Geol. Sci. Hung.*, 11(4): 477–489.

KOMLOSSY, G., 1968. Étude minéralogique et génétique de la pyritisation de quelques bauxites hongroises à l'exemple de l'occurence à Iszkaszentgyörgy. *Symp. Icoba, Zagreb*, 5: 71–81.

KONTA, J., 1958. Proposed classification and terminology of rocks in the series bauxite–clay–iron ore. *J. Sediment. Petrol.*, 28: 83–86.

KÖSTER, H. M., 1961. Vergleich einiger Methoden zur Untersuchung von geochemischen Vorgängen bei der Verwitterung. *Beitr. Mineral. Petrog.*, 8: 69–83.

KRAUSKOPF, K. B. 1956. Dissolution and precipitation of silica at low temperatures. *Geochim. Cosmochim. Acta*, 10: 1–26.

KRAUSKOPF, K. B., 1959. The geochemistry of silica in sedimentary environments. *Soc. Econ. Paleontologists Mineralogists, Spec. Publ.*, 7: 4–19.

KRISHNAN, M. S., 1942. Bauxite in den Shevaroy-Hills, Salem-district Madras residency. *Records Geol. Surv. India*, 77: 1–16.

LACROIX, A., 1901–1909. *Groupe de Diaspore. (Minéralogie de la France et de ses Colonies, 3)* Masson, Paris, 812 pp.

LACROIX, A., 1913. Les latérites de la Guinée et les produits d'altération qui leur sont associés. Nouv. Arch. Mus. V, No. 5, 255–358, Paris.

LACROIX, A., 1923. *Minéralogie de Madagascar*. Paris, 437 pp.

LENGWEILER, H., BUSER, W. and FEITKNECHT, W., 1961. Die Ermittlung der Löslichkeit von Eisen (III) Hydroxiden mit ^{59}Fe. *Helv. Chim. Acta*, 44: 796–811.

LIEBRICH, A., 1892. Beitrag zur Kenntnis der Bauxitez vom Vogelsberg. *Ber. Oberhess. Ges. Natur. Heilk., Giessen, Naturw. Abt.*, 28: 57–98.

LIMA DE FARIA, J., 1963. Dehydration of goethite and diaspore. *Z. Krist.*, 119: 176–203.

LIPPENS, B. C. and DE BOER, J. H., 1964. Study of phase transformations during calcination of aluminium hydroxides by selected area electron diffraction. *Acta Cryst.*, 17: 1312–1321.

LOUGHNAN, F. C. and BAYLISS, P., 1961. The mineralogy of the bauxite deposits near Weipa, Queensland. *Am. Mineralogist*, 46: 209–217.

LUCCA, V., 1966. Contribution à la connaissance de la génèse de certaines bauxites de la R.S. Roumanie. *Bull. Serv. Carte Géol. Alsace Lorraine*, 19: 287–295.

MACK, E., 1964. Berechnung und Schätzung der Bauxitvorräte im Parnass-Kiona-Gebirge. *Berg Hüttenmänn. Monatsh.*, 109: 218–223.

MAIGNIEN, R., 1964. Survey of research on laterites. UNESCO, 125: 1–25.

MAKSIMOVIĆ, Z., 1968. Distribution of trace elements in bauxite deposits of Herzegovina, Yugoslavia. *ICOBA, Zagreb*, 5: 63–70.

MALYAVKIN, W. F., 1926. Bauxite. *Acad. Sci. U.S.S.R., Bull.*, 1926: 145–178.

MARIĆ, L., 1965. *Terra Rossa u Karstu Jugoslavije*. Jugoslav. Akad. Znanosti Umjetnosti, Zagreb, 72 pp.

MARIĆ, L., 1967. Karstifikacija i geokemijska migradija nekojih makroelementata i mikroelementata u jz Dinaridima (Jugoslavija). *Zeml. Biljka*, 16: 539–547.

MARSHALL, C. E., 1940. A petrographic method for the study of soil formation process. *Soil Sci. Soc. Am. Proc.*, 5: 100–103.

MARSHALL, C. E. and HASEMANN, J. F., 1942. The quantitive evaluation of soil formation and development by heavy mineral studies: A grundy silt loam profile. *Soil. Sci. Soc. Am. Proc.*, 7: 448–453.

MATSUSAKA, Y., SHERMAN, G. D. and SWINDALE, L. D., 1965. Nature of magnetic minerals in Hawaiian soils. *Soil Sci.*, 100: 192–199.

McQUEEN, H. S., 1943. Geology of the fire clay districts of east central Missouri. *Missouri Geol. Surv. Water Resources*, 28: 250 pp.

MEGAW, H. D., 1934. The crystal structure of hydrargittite. *Z. Krist.*, 87: 185–204.

MIGDISOW, A. A., 1960. On the titanium-aluminium ratio in sedimentary rocks. *Geochemistry*, 2: 178–194.

MILLOT, G. and BONIFAS, M., 1955. Transformation isovolometriques dans les phénomènes de latéritisation et de bauxitisation. *Bull. Serv. Carte Géol. Alsace Lorraine*, 8(1): 3–19.

MILLOT, G., LUCAS, J. and PAQUET, H., 1966. Evolution géochimique par dégradation et agradation des minéraux argileux dans l'hydrosphère. *Geol. Rundschau*, 55: 1–20.

MOHR, E. C. J., 1938. De bodem der tropen in het algemeen, en die van Nederlandsch-Indië in het bijzonder. *Mededel. Kon. Inst. Tropen, Amsterdam*, 31(2): 1151.

MONTAGNE, D. G., 1964. New facts on the geology of the "young" unconsolidated sediments of northern Surinam. *Geol. Mijnbouw*, 43: 499–515.

MOSES, J. H. and MICHELL, W. D., 1963. Bauxite deposits of Guiana and Surinam. *Econ. Geol.*, 58: 250–262.

MOUSSOULOS, L., 1957. *Le Gisement nickelifère de Larymna, Étude de ses Caractères fondamentaux et de son Méthode d'Exploitation*. Ann. Géol. Pays Helléniques, 9: 1–58. Athens.

MÜCKENHAUSEN, E., 1962. *Entstehung, Eigenschaften und Systematik der Böden der Bundesrepublik Deutschland*. D.L.G.-Verlag, Frankfurt/Main, 148 pp.

NAKAMURA, M. T. and SHERMAN, G. D., 1965. The genesis of halloysite and gibbsite from Mugearite on the Island of Maui. *Hawaii Agr. Expt. Sta., Tech. Bull.*, 62: 1–36.

NEUHAUS, A. and HEIDE, H., 1965. Hydrothermaluntersuchungen im System Al_2O_3–H_2O, 1. Zustandsgrenzen und Stabilitätsverhältnisse von Böhmit, Diaspor und Korund im Druckbereich 50 bar. *Ber. Deut. Keram. Ges.*, 42: 167–184.

NEWBOLD, T. J., 1844. Notes, chiefly geological, across the peninsula from Masuliptam to Goa, comprising remarks on the regur and laterite; occurrence trace of aqueous denudation on the surface of southern India. *J. Asiatic Soc. Bengal*, 13: 984–1004.

NEWSOME, J. W., HEISER, H. W., RUSSELL, A. S., STUMPF, H. C. (Editors), 1960. *Alumina Properties Alcoa Res. Laboratories, Tech. Paper*, 10: 88 pp.

NIA, R., 1968. *Geologische, petrographische, geochemische Untersuchungen zum Problem der Boehmit-Diaspor-Genese in griechischen Oberkreidebauxiten der Parnass-Kiona-Zone*. Thesis Univ. Hamburg, 133 pp.

NICOLAS, J., 1968. Nouvelles données sur la genèse des bauxites à mur karstique du sud-est de la France. *Mineral. Deposita*, 3: 18–33.

NICOLAS, J. and ESTERLE, M., 1965. Position et âge de la bauxite karstique d'Ollière (Var). *Compt. Rend.*, 260: 3722–3723.

NORRISH, K. and TAYLOR, R. M., 1961. The isomorphous replacement of iron by aluminium in soil goethites. *J. Soil Sci.*, 12: 294–306.

OERSTED, H. C., 1824. *Oversight over Videnskabernes Selskabs Forhandlinger*, 5. Kopenhagen, pp.15–16.

OKAMOTO, G., OKURA, T. and GOTO, K., 1957. Properties of silica in water. *Geochim. Cosmochim. Acta*, 12: 123–132.

OVERSTREET, E. C., 1964. Geology of the southeastern bauxite deposits. *U.S. Geol. Surv., Bull.*, 1199-A: 19 pp.

OWEN, H. B., 1954. Bauxite in Australia. *Australia, Bur. Mineral Resources, Geol. Geophys., Bull.*, 24: 234 pp.

PA HO HSU, and BATES, TH. F., 1964. Formation of X-ray amorphous and crystalline aluminium hydroxides. *Mineral. Mag.*, 33(264): 749–768.

PAPASTIAMATIOU, J., 1965. *Quelques Observations sur la Genèse des Bauxites en Grèce, 1*. ICOBA, Zagreb, pp.3–8.

PATTERSON, S. H., 1967. Bauxite reserves and potential aluminium resources of the world. *U.S. Geol. Surv., Bull.*, 1228: 176 pp.

PATTERSON, S. H. and ROBERTSON, C. E., 1961. Weathered basalt in the eastern part of Kauai, Hawaii. *U.S. Geol. Surv., Prof. Papers*, 424-C: 195–198.

PEDRO, G., 1964. Contribution à l'étude expérimentale de l'altération géochimique des roches cristallin. *Ann. Agron.*, 15(2,3,4): 344 pp.

PEDRO, G., 1966. Intérêt géochimique et signification minéralogique du paramètre moléculaire $Ki = SiO_2/Al_2O_3$ dans l'étude des latérites et bauxites. *Bull. Groupe Franç. Argiles*, 18(13): 19–31.

PEDRO, G. and BERRIER, J., 1966. Sur l'altération expérimentale de la kaolinite et sa transformation en boehmite par lessivage à d'eau. *Compt. Rend.*, 262: 729–732.

PEDRO, G. and BITAR, K. E., 1966. Sur l'influence du type chimique de la roche mère dans le développement des phénomènes d'altération superficielle: recherches expérimentales sur l'évolution des roches ultrabasiques (serpentinites). *Compt. Rend.*, 263D: 313–316.

POSNJAK, E. and MERWIN, H. E., 1922a. Hydrated ferric oxides. *Am. J. Sci.*, 47(1919): 311–348.

POSNJAK, E. and MERWIN, H. E., 1922b. The system Fe_2O_3–SO_3–H_2O. *J. Am. Chem. Soc.*, 44: 1965–1994.

PRESCOTT, J. A. and PENDLETON, R. L., 1952. Laterite and lateritic soils. *Commonwealth Bur. Soil Sci., Tech. Commun.*, 47: 51 pp.

RAGLAND, J. L. and COLEMAN, N. T., 1960. The hydrolysis of Al-salts in clay and soil systems. *Soil Sci. Soc. Am. Proc.*, 24: 457–460.

RAN, T. V. M., 1931. Bauxite from Kashmir. *Mineral. Mag.*, 22: 87–91.

RAUPACH, M., 1960. Aluminium ions in aluminium hydroxide, phosphate and soil-water systems. *Nature*, 188: 1049–1050.

RAUPACH, M., 1963. Solubility of simple aluminium compounds expected in soils, 1–III. *Australian J. Soil Res.*, 1(1): 28–35; 36–45; 46–54.

RICHARD, L. W., 1963. The composition and constitution of red and yellow soil earths that may be substituted for bauxite in certain industrial operations. *Econ. Geol.*, 58(1): 138–142.

ROCH, E., 1956. Les bauxites de Provence: des poussières fossiles? *Compt. Rend.*, 242: 2847–2849.

ROCH, E., 1958. Les bauxites de la Midi de la France. *Rev. Gén. Sci. Pures Appl., Bull. Assoc. Franç. Avan. Sci.*, 66(5/6): 151–156.

ROCH, E., 1967. Les problèmes relatifs à la géologie et la biogéographie des terrains renfermant les bauxites et les ocres. *Compt. Rend. Coll. Biostratigr. Crétacé-Eocène France Meridionale*, 21–27.

ROCH, E. and DEICHA, G., 1966. Sur les "argilites" de la région de Draguignan (Var). *Compt. Rend. Soc. Géol. France*, 1966: 145–147.

ROY CHOWDHURY, R., 1958. Bauxite in Bihar, Madhya Pradesh, Vindhya Pradesh, Madhya Bharat and Bopal. *Mem. Geol. Surv. India*, 85: 1–271.

ROY CHOWDHURY, M. K. and ANAND ALWAR, M. A., 1964. Loss on ignition as a guide in prospecting for bauxite, *Indian Minerals*, 18: 49–54.

RUTTNER, A. and WOLETZ, G., 1956. Die Gosau von Weißwasser bei Unterlaussa. *Mitt. Geol. Ges. Wien*, 48: 221–256.

SAALFELD, H., 1960. Strukturen des Hydrargillits und der Zwischenstufen beim Entwässern. *Neues Jahrb. Mineral., Abh.*, 95: 1–87.

SAALFELD, H. and MEHROTRA, B. B., 1965. Elektronenbeugungs-Untersuchungen an Aluminium-oxiden. *Z. Deut. Keram. Ges.*, 42: 161–166.

SAALFELD, H. and MEHROTRA, B. B., 1966. Zur Struktur von Nordstrandit Al $(OH)_3$. *Naturw. Hefte*, 5: 128–129.

SAINTE-CLAIRE DEVILLE, H., 1854. *Ann. Chim. Phys.*, 43: 27; *Compt. Rend.*, 46: 452.

SAKAČ, K., 1966. Marine fossils in the bauxites of Dalmatia. *Geol. Vjesnik*, 19: 131–138.

SAPOZHNIKOV, D. G., 1963. On the subtraction of aluminium by organic acids from minerals and rocks in the course of weathering. *Symp. Bauxites, Zagreb, 1963*, 1: 107–113.

SAWHNEY, B. L., 1958. Aluminium interlayers in soil clay minerals, montmorillonite and vermiculite. *Nature*, 182: 1595–1596.

SCHELLMANN, W., 1964. Zur lateritischen Verwitterung von Serpentinit. *Geol. Jahrb.*, 81: 645–678.

SCHELLMANN, W., 1969. Die Bildungsbedingungen sedimentärer Chamosit- und Hämatit- Eisenerze am Beispiel der Lagerstätte Echte. *Neues Jahrb. Mineral. Abh.*, 111: 1–31.

SCHINDLER, P., MICHAELIS, W. and FEITKNECHT, W., 1963. Löslichkeitsprodukte von Metalloxiden und -hydroxiden: die Löslichkeit gealterter Eisen (III) hydroxidfällungen. *Helv. Chim. Acta*, 46: 444–449.

SCHMALZ, R., 1959. A note on the system Fe_2O_3–H_2O. *J. Geophys. Res.*, 64: 575–579.

SCHROLL, E. and SAUER, D., 1963. Ein Beitrag zur Geochemie der seltenen Elemente in Bauxiten. *Symp. Bauxites,* I. *Zagreb, 1963,* 201–225.

SCHROLL, E. and SAUER, D., 1968. Beiträge zur Geochemie von Titan, Chrom, Nickel, Cobalt, Vanadium und Molybdän in bauxitischen Gesteinen und das Problem der stofflichen Herkunft des Aluminiums. *ICOBA Zagreb,* 5: 83–96.

SCHÜLLER, A., 1957. Mineralogie und Petrographie neuartiger Bauxite aus dem Gun Distrikt, Honan-Provinz (China). *Geologie,* 6: 379–399.

SCHWERTMANN, U., 1959. Über die Synthese definierter Eisenoxide. *Z. Anorg. Allgem. Chem.,* 298: 339–348.

SCHWERTMANN, U., 1966. Die Bildung von Goethit und Hämatit in Böden und Sedimenten. *Proc. Intern. Clay Conf., 1966,* 1: 159–165.

SCHWIERSCH, R., 1933. Thermischer Abbau der natürlichen Hydroxide des Aluminiums und des dreiwertigen Eisens. *Chem. Erde,* 8: 252–315.

SÉGALEN, P., 1965. Les produits alumineux dans les sols de la zone tropicale humide, 1. 2. *ORSTOM Cahiers, Pédol.,* 3: 149–176; 179–205.

SHAPIRO, L. and BRANNOCK, W. W., 1956. Rapid analysis of silicate rocks. *U.S. Geol. Surv. Bull.,* 1036 (C): 53–84.

SHERMAN, G. D., 1952. The titanium content of Hawaiian soils and its significance. *Soil Sci., Proc.,* 16: 15–18.

SHERMAN, G. D., 1958. Gibbsite-rich soils of the Hawaiian Islands. *Hawaii Agr. Expt. Sta., Bull.,* 116: 1–23.

SHERMAN, G. D. and UEHARA, G., 1956. The weathering of Olivine Basalt in Hawaii and its Pedogenic Significance. *Soil Sci. Soc. Am., Proc.,* 20: 337–340.

SHERMAN, G. D., IKAWA, H., UEHARA, G. and OKAZAKI, E., 1962. Types of occurrence of nontronite and nontronite-like minerals in soils. *Pacific Sci.,* 16: 57–62.

SHERMAN, G. D., MATSUSAKA, Y., IKAWA, H. and UEHARA, G., 1964. The role of the amorphous fraction in the properties of tropical soils. *Agrochimica,* 8: 146–163.

SINCLAIR, J. G. L., 1967. Bauxite genesis in Jamaica: new evidence from trace element distribution. *Econ. Geol.,* 62: 482–486.

SIVARAJASINGHAM, S., ALEXANDER, L. T., CANDY, J. G. and CLINE, M. G., 1962. Laterite. *Advan. Agron.,* 14: 1–60.

SLANSKY, M., LALLEMAND, A. and MILLOT, G., 1964. La sédimentation et l'alteration laterique des formations phosphatées du gisement de Taiba, République de Sénégal. *Bull. Serv. Carte Géol. Alsace Lorraine,* 17: 311–324.

SMITH, G. F. and KIDD, D. J., 1949. Hematite-goethite relations in neutral and alkaline solutions under pressure. *Am. Mineralogist,* 34: 403–412.

SMITH-BRACEWELL, S., 1962. *Bauxite, Alumina and Aluminium.* H.M. Stationary Office, London, 234 pp.

STRACHOW, M. M., 1961. *Grundzüge der Lithogenetischen Theorie, 1.* Nauka, Moscow, 2nd ed., 212 pp. (in Russian).

STRENG, A., 1858. Über den Melaphyr des südlichen Harzrandes. *Z. Deut. Geol. Ges.,* 10: 99.

STRENG, A., 1860. Die quarzführenden Porphyre des Harzes. *Neues Jahrb. Mineral. Geol. Palaeontol.,* 129: 218.

STRUNZ, H., 1966. *Mineralogische Tabellen.* Akad. Verlagsgesellschaft, Leipzig, 560 pp.

SZABÓ, P. Z., 1964. Neue Daten und Beobachtungen zur Kenntnis der Paläokarsterscheinungen in Ungarn. *Erdkunde,* 18: 135–142.

SZANTNER, F. and SZABÓ, E., 1962. Uj tektonikai megfigyelések az utobbiévek bauxit Kutatásái alapian. *Földt. Közl.,* 92: 416–451.

TAMURA, T., 1958. Identification of clay minerals from acid soils. *J. Soil Sci.,* 9: 141–147.

TANADA, T., 1959. Certain properties of the inorganic colloidal fraction of the Hawaiian soils. *J. Soil. Sci.,* 2: 83–96.

TERENTIEVA, K. F., 1958. La genèse des minéraux de l'alumine dans la bauxite. In: *Bauxites, Minéralogie et Genèse*. Izdat. Akad. Nauk. SSSR, Moscow, (in Russian).

TESSIER, F., 1965. Les niveaux latéritiques du Sénégal. *Ann. Fac. Sci. Univ. Marseille*, 37: 221–237.

THIEL, R., 1963. Zum System α-FeOOH–α-ALOOH. *Z. Anorg. Allgem. Chem.*, 326: 70–78.

TORKAR, K., 1960. Untersuchungen über Aluminium-hydroxide und -oxide. *Monatsh. Chemie.*, 91: 400–450; 653–658; 757–764.

TORKAR, K. and KRISCHNER, H., 1963. Über die Eigenschaften reiner Aluminiumhydroxyde und Oxyde. erhalten mit hydrothermalen Darstellungsmethoden. *Symp. Bauxites, Zagreb, 1963*: 25–35.

TRÖGER, W. E., 1952. *Tabellen zur optischen Bestimmung der gesteinsbildenden Minerale*. Schweizerbart, Stuttgart, 147 pp.

TUCAN, F., 1912. Terra rossa, deren Natur und Entstehung. *Neues Jahrb. Mineral. Geol. Palaeontol.*, 34: 401–430.

TUREKIA, K. K. and WEDEPOHL, H. K., 1961. Distribution of the elements in some major units of the earth's crust. *Bull. Geol. Soc. Am.*, 72: 175–192.

UEHARA, G., IKAWA, H. and SHERMAN, G. D., 1966. Desilication of halloysite and its relation to gibbsite formation. *Pacific Sci.*, 20: 119–124.

U.S. SOIL SURVEY STAFF, 1960. *Soil Classification, a Comprehensive System, 7th Approximation*. U.S. Govt. Printing Office, Washington, D. C., 1265 pp.

VADAZ, E., 1943. Alunit a magyar bauxitelöfordulá sokban. *Föld. Közl.*, 73: 169–179.

VADAZ, E., 1951. *Bauxitföldtan*. Akad. Kiado, Budapest, 129 pp.

VALETON, I., 1964. Problems of boehmitic and diasporitic bauxites. In: G. C. AMSTUTZ (Editor), *Sedimentology and Ore Genesis*, 2. Elsevier, Amsterdam, pp.123–129.

VALETON, I., 1965. Faziesprobleme in südfranzösischen Bauxitlagerstätten. *Beitr. Mineral. Petrog.*, 11: 217–246.

VALETON, I., 1966. Sur la genèse des gisements de bauxite du sud-est de la France. *Bull. Soc. Géol. France*, 7: 685–701.

VALETON, I., 1967a. Laterite und ihre Lagerstätten. *Fortschr. Mineral.*, 44: 67–130.

VALETON, I., 1967b. Einige optische und chemische Eigenschaften indischer Gibbsite. *Johanneum, Mineral. Mitt.*, 1–2: 113–123.

VALETON, I., 1967c. Bauxitführende Laterite auf dem Trappbasalt Indiens als fossile, polygenetisch veränderte Bodenbildung. *Sediment. Geol.* 1: 7–56.

VALETON, I., 1968. Zur Petrographie der Bauxitlagerstätten auf der "Charnockite-suite" im Salemdistrikt und in den Nilgiri-Hills, Südindien. *Mineral. Deposita (Berlin)* 3: 34–47.

VAN BEMMELEN, R. W., 1941. Origin and mining of bauxite in Netherlands India. *Econ. Geol.*, 36: 630–640.

VAN DER HAMMEN, T. and WYMSTRA, T. A., 1964. A palynological study on the Tertiary and Upper Cretaceous of British Guiana. *Leidse Geol. Mededel.*, 30: 184–240.

VON SIEMENS, W., 1878. Engl. Patent 2110, 1879.

WALTHER, J., 1915. Laterite in West-Australien. *Z. Deut. Geol. Ges.*, *Monatsber.*, 67: 113–140.

WALL, J. R. D., WOLFENDEN, E. B., BEARD, E. H. and DEANS, T., 1962. Nordstrandite in soil from West Sarawak, Borneo. *Nature*, 4851: 264–266.

WEFERS, K., 1962. Zur Struktur der Aluminiumtrihydroxide. *Naturwissenschaften*, 49: 204–205.

WEFERS, K., 1966. Zum System Fe_2O_3–H_2O. *Ber. Deut. Keram. Ges.*, 43: 677–684; 703–708.

WEFERS, K., 1967. Phasenbeziehung im System Al_2O_3–Fe_2O_3–H_2O. *Erzmetall*, 20: 13–19; 71–75.

WEINMANN, B., 1964. Die Böden der Insel Kefallina. *Giessener Abhandl. Agrar. Wirtsch. forsch.*, 28: 213 pp.

WEY, R. and SIEFERT, B., 1961. Réaction de la silice monomoléculaire en solution avec les ions Al^{3+} et Mg^{2+}. Génèse et synthèse des argiles. *Colloq. Intern. Centre Natl. Rech. Sci. (Paris)*, 105: 11–23.

WIPPERN, J., 1965. Die Ausgangsgesteine für Bauxitbildung. *Bull. Mineral Res. Exploration Inst. Turkey*, Foreign Ed., 60: 40–44.

WIPPERN, J., 1967. Die Aluminiumrohstoffe der Türkei. *Bull. Mineral Res. Exploration Inst. Turkey*, 62: 83–90.

WOLFENDEN, E. B., 1961. Bauxite in Sarawak. *Econ. Geol.*, 56: 972–981.

WOLFENDEN, E. B., 1965. Geochemical behaviour of trace elements during bauxite formation in Sarawak, Malaysia. *Geochim. Cosmochim. Acta*, 29: 1051–1062.

WOLLAST, R., 1963. Aspect chimique du monde de formation des bauxites dans le Bas-Congo. Confrontation des données thermodynamiques et expérimentales. *Acad. Roy. Sci. Outre-Mer, Bull. Belg., Sér. 7*, 2: 392–412.

WORLD SOIL RESOURCES REPORTS. Meeting on the classification and correlation of soils from volcanic ash, Tokio, 1964. *World Soil Res. Rept.*, 1964: 139–142.

WYART, J., OBERLIN, A. and TCHOUBAR, C., 1966. Étude en microscopie et microdiffraction électroniques de la boehmite formée lors de l'altération de l'albite. *Centre Natl. Rech. Sci., France, Bull.*, 18(14): 51–57.

ZAMBO, J. and TOTH, P., 1961. Magyar bauxitok feltárhátoságarol. *Femip. Kut. Int. Közlemén.*, 5: 19–27.

ZANS, V. A., 1951. On karst hydrology in Jamaica. *Union Géol. Geophys. Intern. Assoc. Intern. Hydrol. Sci.*. 2: 267–279.

ZANS, V. A., 1954. Bauxite resources of Jamaica and their development. *Geol. Surv., Dept. Jamaica, West Indies Bull.*, 1: 307–332.

ZANS, V. A., 1959. Recent views on the origin of Bauxite. *Geonotes*, 1(5): 123–132.

ZAPP, A.D., 1965. Bauxite deposits of the Andersonville district, Georgia. *U.S. Geol. Surv., Bull.*, 1199-G: 37 pp.

ZIA-UL HASAN, 1966. On the occurrence and geochemistry of bauxite deposits of Monghyr area (India). *Econ. Geol.*, 61: 715–730.

Index